COLORADO GEOLOGICAL SURVEY RESOURCE SERIES 30
BUREAU OF ECONOMIC GEOLOGY REPORT OF INVESTIGATIONS NO.220

Geologic and Hydrologic Controls on Coalbed Methane: Sand Wash Basin, Colorado and Wyoming

By
W. R. Kaiser, A. R. Scott, D. S. Hamilton, Roger Tyler,
R. G. McMurry, and Naijiang Zhou
Bureau of Economic Geology
The University of Texas at Austin
Austin, TX 78713-7508

and

C. M. Tremain
Colorado Geological Survey

Sponsored by
Gas Research Institute
Contract No. 5091-214-2261
GRI Topical Report No. 92/0420
1993

Colorado Geological Survey
Division of Minerals and Geology
Department of Natural Resources
Denver, Colorado
1994

Cover Photo: Twentymile Sandstone outcrop in Routt County, Colorado, above Cyprus Coal Company's Foidel Creek Mine

50272-101

REPORT DOCUMENTATION PAGE

1. REPORT NO. GRI-92/0420	**2.**	**3. Recipient's Accession No.**	

4. Title and Subtitle

Geologic and Hydrologic Controls on Coalbed Methane: Sand Wash Basin, Colorado and Wyoming

5. Report Date August 1993

6.

7. Author(s) W. R. Kaiser, A. R. Scott, D. S. Hamilton, Roger Tyler, R. G. McMurry, Naijiang Zhou, and C. M. Tremain

8. Performing Organization Rept. No.

9. Performing Organization Name and Address

Bureau of Economic Geology
The University of Texas at Austin
University Station, Box X
Austin, TX 78713-7508

10. Project/Task/Work Unit No.

11. Contract(C) or Grant(G) No.

(C) 5091-214-2261

(G)

12. Sponsoring Organization Name and Address

Gas Research Institute
8600 West Bryn Mawr Avenue
Chicago, IL 60631 Project Manager: Richard A. McBane

13. Type of Report & Period Covered

Topical Report
8/1/91–4/30/93

14.

15. Supplementary Notes

16. Abstract (Limit: 200 words)

Coal resources occur mainly in the Upper Cretaceous Williams Fork Formation and Paleocene Fort Union Formation in the eastern part of the Sand Wash Basin. These coals are mostly subbituminous to high-volatile B bituminous rank and have average gas contents of less than 200 ft^3/ton (<6.24 m^3/t). Coalbed methane resources total 101 Tcf (2.86 Tm^3). The basin's cumulative gas/water ratio is approximately 15 ft^3/bbl (~2.7 m^3/m^3). To date, low gas content and high water production have limited coalbed methane activity in the basin. Steep structural dip and coal distribution have restricted exploration to the eastern and southeastern margins of the basin. Prospective Williams Fork and Fort Union coals lie basinward, northwest of Craig, Colorado, in association with the Cedar Mountain fault system and westward along the Cherokee Arch, respectively. High productivity requires that permeability, ground-water flow direction, coal distribution and rank, gas content, and structural grain be synergistically combined. That synergism explains prolific and marginal production in the San Juan and Sands Wash Basins, respectively. On the basis of a comparison between the basins, a basin-scale coalbed methane producibility model is proposed. High productivity is thought to require ground-water flow through coals of high gas content orthogonally toward no-flow boundaries and conventional trapping of gas along them.

17. Document Analysis a. Descriptors

Coalbed methane, Sand Wash Basin, Colorado, Upper Cretaceous, Mesaverde Group, Paleocene, Williams Fork Formation, Fort Union Formation, structural setting, cleat, depositional systems, genetic stratigraphy, coal occurrence, coal rank, gas content, hydrology, hydrochemistry, resources, production

b. Identifiers/Open-Ended Terms

Methane from coal beds, occurrence and production, geologic and hydrologic controls, gas and coal resources, producibility model

c. COSATI Field/Group

18. Availability Statement	**19. Security Class (This Report)**	**21. No. of Pages**
Release unlimited	Unclassified	151
	20. Security Class (This Page)	**22. Price**
	Unclassified	

See ANSI—Z39.18) *See Instructions on Reverse* OPTIONAL FORM 272 (4–77) (Formerly NTIS–35) Department of Commerce

Disclaimer

Research Summary

Title

Geologic and Hydrologic Controls on Coalbed Methane: Sand Wash Basin, Colorado and Wyoming

Contractor

Bureau of Economic Geology, The University of Texas at Austin, GRI Contract No. 5091-214-2261

Principal Investigator

W. R. Kaiser

Report Period

August 1, 1991–April 30, 1993

Objectives

To identify geologic and hydrologic controls on the occurrence and producibility of coalbed methane in the Sand Wash Basin, northwest Colorado and southwest Wyoming.

Technical Perspective

Coalbed methane production has been established in the Sand Wash Basin. Large coal resources, gas shows during drilling of coal beds, and high gas contents in some coals triggered initial development along the basin margins. Results to date have been disappointing. Coalbed wells have yielded little gas and large volumes of water. In the absence of a regional analysis, neither production data nor the basin's ultimate coalbed methane potential could be fully evaluated. Thus, the need arose for an integrated geologic and hydrologic study of the basin to provide the framework for evaluating development properties and the rationale for future exploration.

Results

Large coal resources occur in the Upper Cretaceous Williams Fork Formation and lower Tertiary Fort Union Formation in the eastern part of the Sand Wash Basin. These coals are mainly subbituminous to high-volatile B bituminous and have average gas contents of less than 200 ft^3/ton. (<6.24 m^3/t). Coalbed methane resources total 101 Tcf (2.86 Tm^3) and are 24 Tcf (680 Bm^3) at shallow drilling depths of less than 6,000 ft (<1,830 m). More than 87 percent of them are in the Williams Fork. The basin's cumulative gas/water ratio is approximately 15 ft^3/bbl (2.7 m^3/m^3). To date, low gas content and high water production have limited coalbed methane activity in the basin. Steep structural dip and coal distribution have restricted exploration to the eastern margins of the basin. Prospective Williams Fork and Fort Union coals, respectively, lie basinward in association with the Cedar Mountain fault system and westward along Cherokee Arch into the Powder Wash field area. High productivity requires that permeability, ground-water flow direction, coal distribution and rank, gas content, and structural grain be synergistically combined. That synergism explains prolific and marginal production in the San Juan and Sand Wash Basins, respectively. On the basis of a comparison between the basins, a basin-scale coalbed methane producibility model is proposed whose essential elements are: ground-water flow through thick coals of high rank and high gas content orthogonally toward no-flow boundaries and conventional trapping of gas along them.

Technical Approach

In geologic studies, approximately 160 and 165 geophysical logs, respectively, were used to evaluate Williams Fork and Fort Union structure, genetic stratigraphy, sedimentology, and coal occurrence. A grid of interlocking cross sections was made to identify and define the major coal-bearing stratigraphic units. Structure-contour maps were made on those units. Major structural elements were further defined from 115 miles of seismic data.

The Williams Fork Formation was divided into four genetic stratigraphic units and the Fort Union into four operationally defined lithostratigraphic units. Lithofacies and coal-occurrence maps were made for each unit. Genetic units provided the foundation for subsurface correlation and mapping and more importantly the basis for predicting the geometry and distribution of framework sandstones and coal deposits in areas of meager control. In the absence of porosity logs, coals were operationally identified by very high resistivity, low natural gamma response, and shale-like SP response. Individual coal beds were correlated on the basis of their gamma-ray and density profiles, seam signatures sensitive to minor fluctuations in the coal lithotypes.

A Mesaverde coal-rank map was made from 50 measured vitrinite reflectance (VR) values from 10 wells, 39 VR values calculated from proximate and ultimate analyses, and 55 VR values calculated from a VR profile. A Fort Union map was made from 40 VR values from 15 wells. Coal heating value (Btu/lb) was converted to equivalent VR. In the absence of measured values and analyses, VR values were calculated from equations established by regression analysis of Mesaverde coal and shale data taken from profiles in the Sand Wash and Washakie Basins. The Fort Union VR profile was established from Fort Union and Mesaverde data. Mesaverde and Fort Union gas-content data (about 250 and 125 values) were obtained from the literature and operators.

Mesaverde and Fort Union hydrology were evaluated in an analysis of hydraulic head, pressure regime, and hydrochemistry. Hydraulic heads were calculated from SIP's recorded in DST's and BHP's calculated from WHSIP's. Approximately 90 Mesaverde and 200 Fort Union head values were used to prepare potentiometric-surface maps. Pressure regime and vertical flow direction were evaluated from simple and vertical pressure gradients, respectively, calculated on data screened from several hundred DST's. Chlorinity and TDS maps, made from 155 water analyses from 66 Mesaverde wells and 136 analyses from 69 Fort Union wells, were used to further evaluate ground-water flow.

Gas and coal resources were calculated from digitized structure, topographic, and net-coal-thickness maps on a 3.5-mi^2 (9.1-km^2) grid, using plots of gas content versus depth, density, and coal volume. Production data were obtained from commercial companies, public agencies, the literature, and operators and were related to the geology and hydrology to identify controls on production.

Project Implications

Geologic and hydrologic controls on the occurrence and producibility of coalbed methane identified in the San Juan Basin under a previous contract (no. 5087-214-1544) were further delineated in this integrated study of the Sand Wash Basin. High productivity requires that these controls be synergistically combined. The proposed basin-scale coalbed methane producibility model provides a rationale for future exploration and development in the Sand Wash Basin and, upon further refinement and testing, other United States coal basins.

Richard A. McBane
GRI Project Manager

CONTENTS

FIGURES

TABLES

UPPER CRETACEOUS MESAVERDE GROUP

STRATIGRAPHY AND COAL OCCURRENCE OF THE UPPER CRETACEOUS MESAVERDE GROUP, SAND WASH BASIN

FIGURES

FIGURES

FIGURES

PALEOCENE FORT UNION FORMATION

STRATIGRAPHY AND COAL OCCURRENCE OF THE PALEOCENE FORT UNION FORMATION, SAND WASH BASIN

FIGURES

FIGURES

FIGURES

RESOURCES AND PRODUCTION

RESOURCES AND PRODUCIBILITY OF COALBED METHANE IN THE SAND WASH BASIN

FIGURES

TABLES

Structural and Stratigraphic Setting

Tectonic Evolution, Stratigraphic Setting, and Coal Fracture Patterns of the Sand Wash Basin

Roger Tyler and C. M. Tremain

Abstract

The principal tectonic features within the Sand Wash Basin of northwest Colorado and southwest Wyoming are the result of latest Cretaceous to earliest Oligocene Laramide deformation. Tectonism has affected depositional patterns, coal occurrence, hydrology, and thermal maturity (gas generation) in the Sand Wash Basin. Tectonism has also determined the distribution and orientation of faults, folds, and fractures within the basin and has produced a predominantly northwest structural grain. Permeability in coals and adjacent rocks is largely due to the occurrence of fractures (cleats) and faults.

Northwest-trending systematic fractures (face cleats) and faults on the southeast margin of the Sand Wash Basin are generally parallel to current maximum horizontal stress directions and may provide permeable pathways for both gas and water; fracture swarms may further enhance coal permeability. This section provides the tectonic and stratigraphic setting for studies of depositional, coal occurrence, hydrologic, and thermal maturity patterns in Cretaceous and Tertiary coal zones of the Sand Wash Basin and includes a summary of observations of fracture patterns and stress regimes. An understanding of the tectonic setting of the basin, combined with the studies in the following chapters, provides a basis for predicting coalbed methane occurrence and producibility.

Location

The Sand Wash Basin of northwest Colorado and southwest Wyoming is a subbasin of the Greater Green River Basin, which is located east of the Wyoming–Idaho segment of the Cordilleran thrust belt (fig. 1). The subbasins are generally delineated by uplifts that resulted from latest Cretaceous to earliest Oligocene Laramide deformation (Baars and others, 1988). The Sand Wash Basin is located in the southeastern part of the Greater Green River Basin and is essentially a southerly extension of the larger Washakie Basin of southern Wyoming; their synclinal axes trend north-south (fig. 1). The east-west-trending Cherokee Arch (ridge), a complexly faulted, westward-plunging anticline (Masters, 1961), separates the Sand Wash Basin from the Washakie Basin. To the east, the Sand Wash Basin is bounded by the Sierra Madre and Park Uplifts, to the south by the White River Uplift, to the southwest by the Uinta Uplift and its southeast extension, the Axial Arch, and to the northwest by the Rock Springs Uplift (fig. 1). The Vermillion Basin area (between T12N, R100W and T13N, R102W), a structural and topographic subbasin between the Rock Springs and Uinta Uplifts (fig. 1), differs from the rest of the Sand Wash Basin in that it has rapid structural and facies changes, and stratigraphic thickness variations (Colson, 1969).

In the Sand Wash Basin, basement rocks are as deep as 17,000 ft (5,182 m) below sea level (Tweto, 1975), and Cambrian through Tertiary-age rocks may be as much as 30,000 ft (9,144 m) thick (Irwin, 1986). During the Upper Cretaceous, as much as 11,000 ft (3,353 m) of clastic sediments were deposited (Haun and Weimer, 1960); Paleocene and Eocene rocks are at least 10,000 ft (3,048 m) thick against the Sierra Madre–Park and Uinta Uplifts. Upper Cretaceous and early Tertiary strata, comprising the Mesaverde Group, Lewis Shale, Fox Hills Sandstone, and Lance and Fort Union Formations (fig. 2), crop out mainly on the eastern and southeastern margins of the basin. On the southeastern side, the strata crop out in a north-south-trending belt for about 50 to 60 mi (80 to 97 km), and in a west-east-trending belt for about 40 to 50 mi (64 to 80 km). The strata dip moderately to steeply basinward, with dips ranging from about 5° to 20°. A small area (6 mi [10 km] long and 2 mi [3 km] wide) of Mesaverde Group strata is exposed on the southwestern flank of the Sand Wash Basin adjacent to the Uinta Uplift.

Tectonic Evolution

During Cretaceous time, the area of the present Sand Wash Basin was near the western margin of the Western Interior Seaway, a shallow sea that extended from north to south across much of the North American Midcontinent (Kauffman, 1977) (fig. 3). The Western Interior Seaway occupied a foreland basin bounded on the west by the Cordilleran thrust belt. Greatest subsidence and deposition was along the western margin of the seaway,

In **Kaiser, W. R., and others, 1993, Geologic and hydrologic controls on coalbed methane: Sand Wash Basin: The University of Texas at Austin, Bureau of Economic Geology, topical report prepared for the Gas Research Institute under contract no. 5091-214-2261 (GRI–92/0420), p. 3–19.**

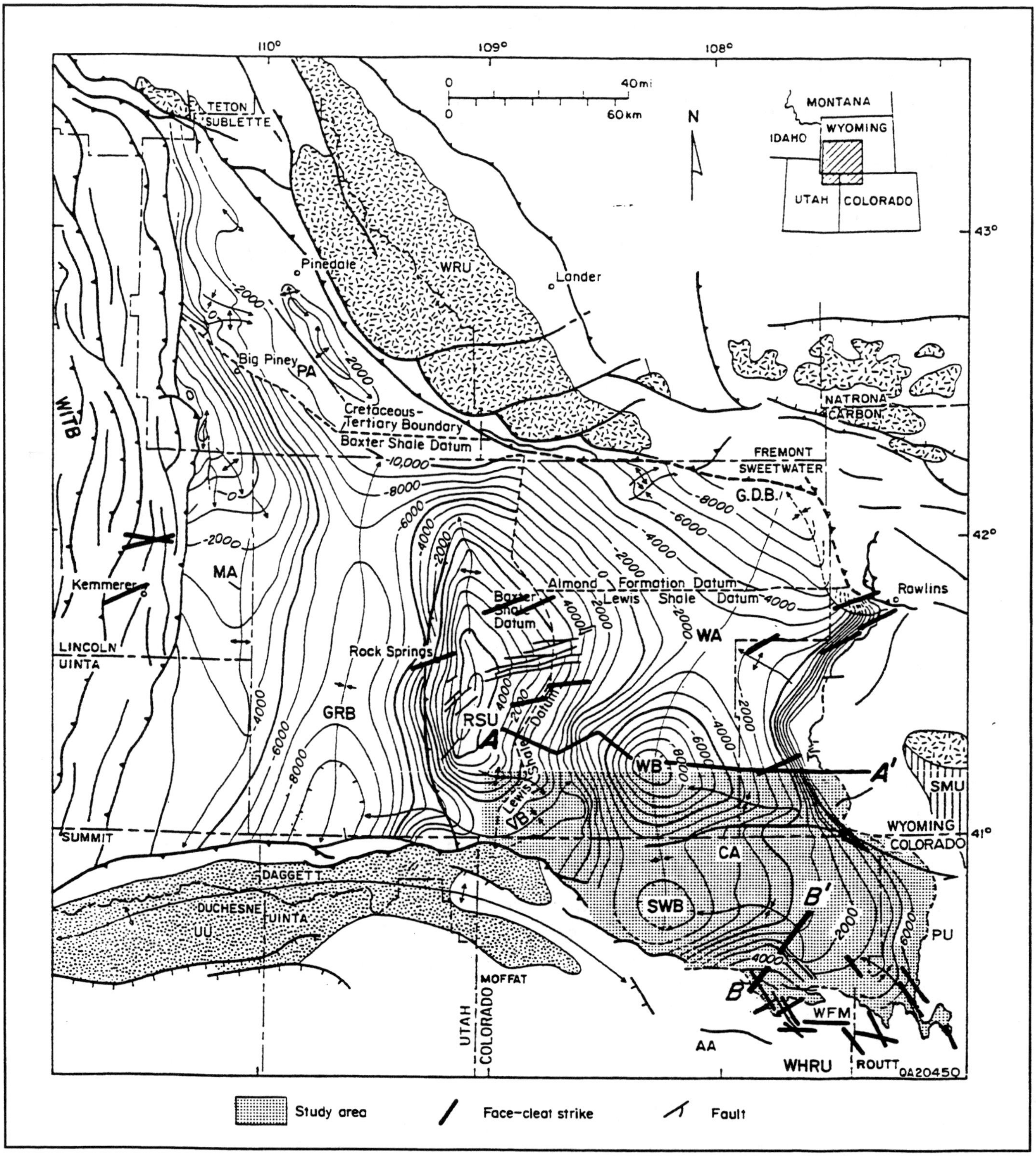

Figure 1. Tectonic map of southwestern Wyoming and adjacent states showing the major tectonic elements of the Greater Green River Basin and the location of the Sand Wash Basin. The map is taken from Lickus and Law (1988). Structure contours, in feet relative to mean sea level, are drawn on Upper Cretaceous marker horizons. Major tectonic features are identified as follows: AA, Axial Arch; CA, Cherokee Arch; GDB, Great Divide Basin; GRB, Green River Basin; PA, Pinedale Anticline; PU, Park Uplift; RSU, Rock Springs Uplift; SMU, Sierra Madre Uplift; SWB, Sand Wash Basin; UT, Uinta thrust fault; UU, Uinta Uplift; WA, Wamsutter Arch; WB, Washakie Basin; WFM, Williams Fork Mountains; WHRU, White River Uplift; WITB, Wyoming–Idaho thrust belt; WRU, Wind River Uplift; and VB, Vermillion Basin. Basement rocks are identified as random-dash, vertical-line, and stippled patterns.

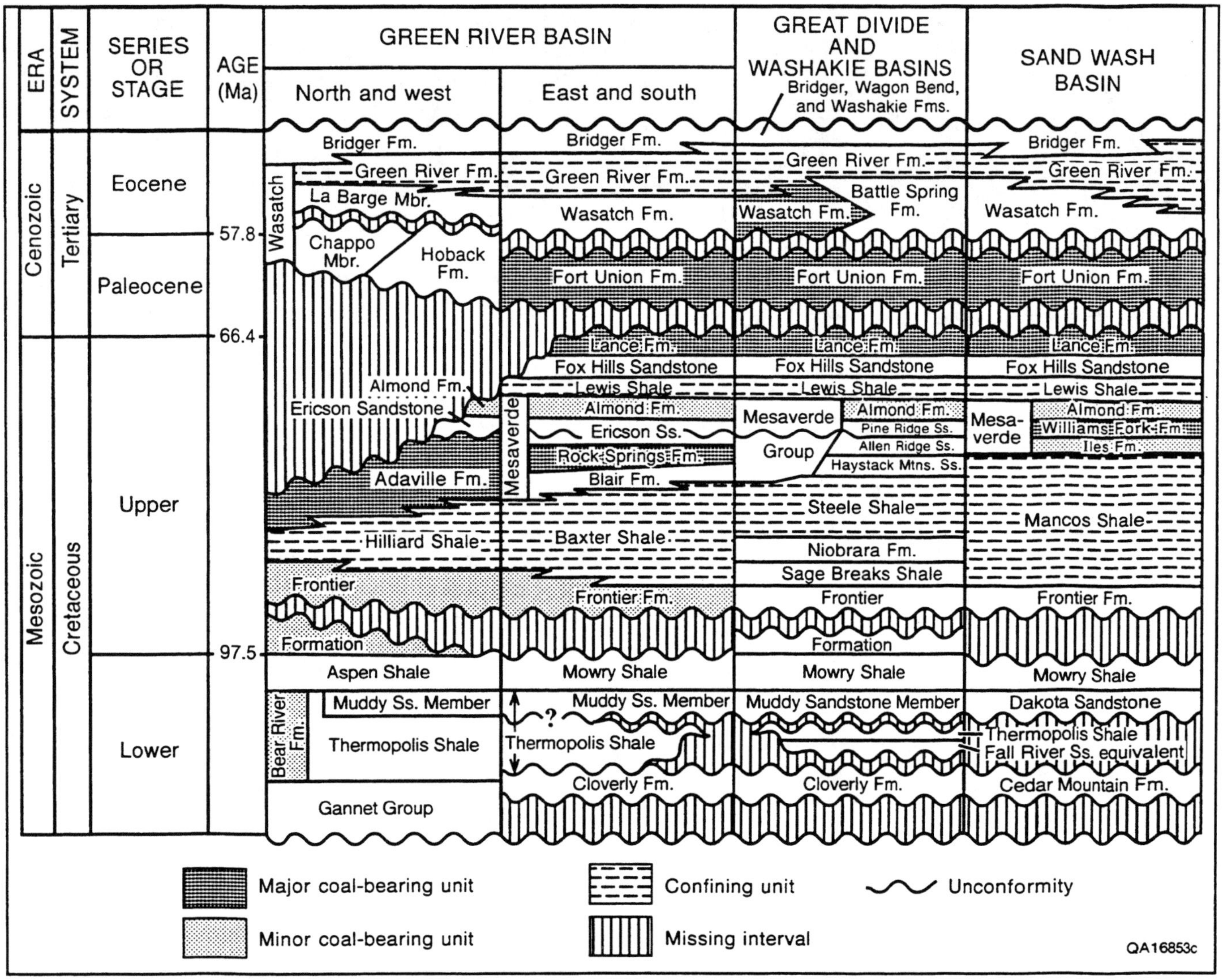

Figure 2. Coal-bearing stratigraphic and confining units in the Sand Wash Basin and surrounding subbasins of the Greater Green River Basin. Modified from Baars and others (1988).

adjacent to the Cordilleran thrust belt. The initiation of deformation in the thrust belt during the Early to Late Cretaceous Sevier Orogeny coincided with a major episode of subsidence of the Western Interior Seaway (Heller and others, 1986). Sediments derived from the uplifts to the west gradually filled the basin, causing the northeast-trending shoreline to retreat eastward. Numerous transgressions and regressions of the shoreline are recorded in the Cretaceous sediments and reflect episodic thrust belt deformation and eustatic sea-level change. The Fox Hills Sandstone (fig. 2) represents the final regressive shoreline facies of the Western Interior Seaway and the Lance Formation the succeeding aggradational facies (Irwin, 1986), terminating Cretaceous sedimentation. The Fox Hills/Lance couplet is depositionally equivalent and homotaxial to the Pictured Cliffs/Fruitland couplet, a prolific gas producer in the San Juan Basin.

In Late Cretaceous to early Miocene time, the Laramide Orogeny caused major uplifts, folds, and faults to propagate in the foreland of the Cordilleran thrust belt. This structural event subdivided the foreland area into individual basins and subbasins, such as the Greater Green River and Sand Wash Basins, respectively (fig. 1). During the Laramide Orogeny, the Sand Wash Basin was filled with fluvial-lacustrine sediments of the Fort Union and Wasatch Formations (fig. 2). The Fort Union and Wasatch Formations contain sediment shed from the surrounding Sawatch Range (Beaumont, 1979; Tyler, this vol.), and the Sierra Madre–Park and Uinta Uplifts (Osmond, 1986; Tyler, this vol.).

Precise timing of the uplifts remains controversial, but preexisting structural grain may have controlled the orientation of some uplifts. For example, structural grain having east-west trends in 2.7-b.y.-old gneisses and quartzites plus seismic data indicate that the Uinta

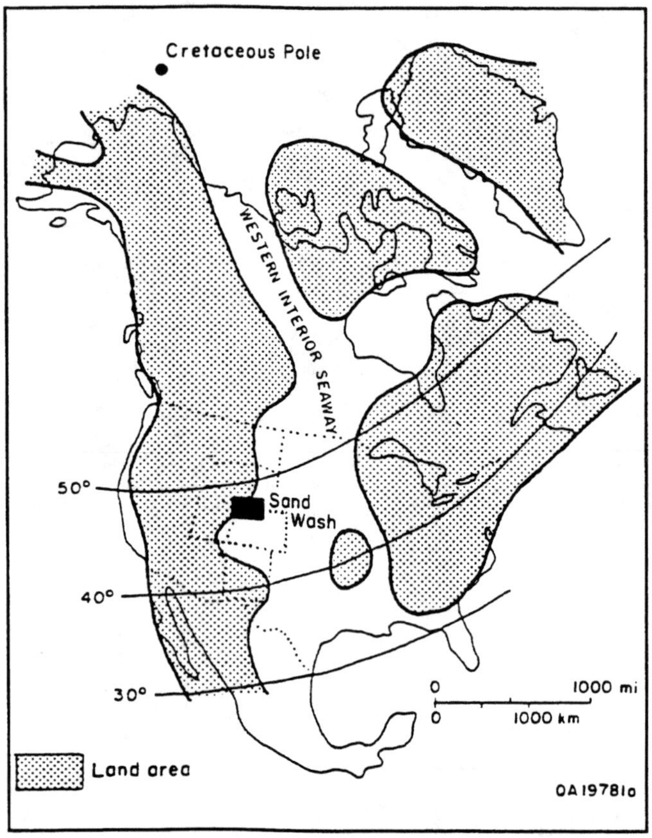

Figure 3. Location of the Sand Wash Basin relative to the Western Interior Seaway. Modified from Kauffman (1977).

Mountains and their southeast extension, the Axial Arch, may have been influenced by faulting dating back to the Proterozoic (Stone, 1975; Hansen, 1986). In addition, the east-west trend of the Cherokee Arch on the northern boundary of the Sand Wash Basin may also have been inherited from major Precambrian structures (Osmond, 1986). Uplift occurred again during the Oligocene, and extensional deformation began in the early Miocene (Hansen, 1986). Extensional faulting continued at a diminished rate into the Quaternary (Hansen, 1986). Dikes, sills, and other intrusives were also emplaced during the late Tertiary (Tweto, 1979) in the eastern part of the basin and locally coked or metamorphosed coals to anthracite (Bass and others, 1955). The dikes exhibit northwesterly trends similar to fractures and faults in the area.

Structure and tectonic maps of the Mesaverde Group and the Fort Union Formation (figs. 4 through 7) indicate that thrust, reverse, and normal fault systems and tight anticlinal folds abound within the structurally complex Sand Wash Basin. The fault systems generally strike northwesterly in the southern portion of the basin but are dominantly westerly striking in the northern part of the basin, with some major faults having a north-northwesterly structural grain. Left lateral strike-slip or wrench faults also occur along the northeastern and southwestern flanks of the Sand Wash Basin (T11N, R85W to R88W and T7N, R94W, respectively). Reverse and/or thrust faults occur on at least one or more sides of the major Laramide uplifts. Vertical separation across the fault systems may exceed 10,000 ft (3,048 m). Faults projected from the subsurface to outcrop parallel the trace of the Mesaverde Group outcrop in the southern and eastern Sand Wash Basin, and the outcrop of the Mesaverde Group and Fort Union Formation in the northern and northeastern Sand Wash Basin, respectively. Major fold axes essentially trend north and northwest in the eastern and central Sand Wash Basin parallel to the strikes of the major fault systems. Minor fold axes near the eastern edge of the basin also have a northeasterly strike component.

Stratigraphic Setting

The Sand Wash Basin is a structurally complex intermontane basin containing three major progradational cycles in Upper Cretaceous strata (fig. 2). The cycles were initiated by tectonic uplift and loading of the Cordilleran thrust belt and eustatic sea-level fluctuations. Each cycle extended deltaic and coastal-plain deposits farther basinward than did the preceding cycle, indicating an overall filling of the Western Interior Seaway. Progradation extended coal-bearing strata (Frontier Formation) (fig. 2) to the east of the Rock Springs Uplift (fig. 1) during the first cycle. Equivalent strata basinward are mud-rich prodelta and delta-front facies. The second major cycle established coal-forming conditions in deltaic and back-barrier settings (Mesaverde Group) beyond the present-day eastern limit of the Sand Wash Basin. Regressive and transgressive cycles are recognized within the major Mesaverde Group cycle. The Fox Hills Sandstone represents the last Cretaceous progradational event in the foreland basin and is the platform upon which Lance Formation coals accumulated (fig. 2).

Basement uplifts subsequently broke the foreland basin into smaller structural and depositional basins during the Laramide Orogeny. Fluvial sandstone and conglomeratic sandstone and floodplain shale, siltstone, and coal are the major lithologic components of the Paleocene Fort Union Formation. Early Eocene time brought an even greater period of crustal instability to the region. The Fort Union Formation was uplifted throughout the region, tilted and truncated along the margins of the basement uplift, and covered by sandstone and variegated shale of the Wasatch Formation (McDonald, 1972, 1975; Tyler, this vol.). By middle Eocene time, structural and topographic relief had developed to the extent that the Sand Wash Basin probably became a closed topographic basin and contained an extensive lacustrine system. Following the Laramide Orogeny an extensional stress regime,

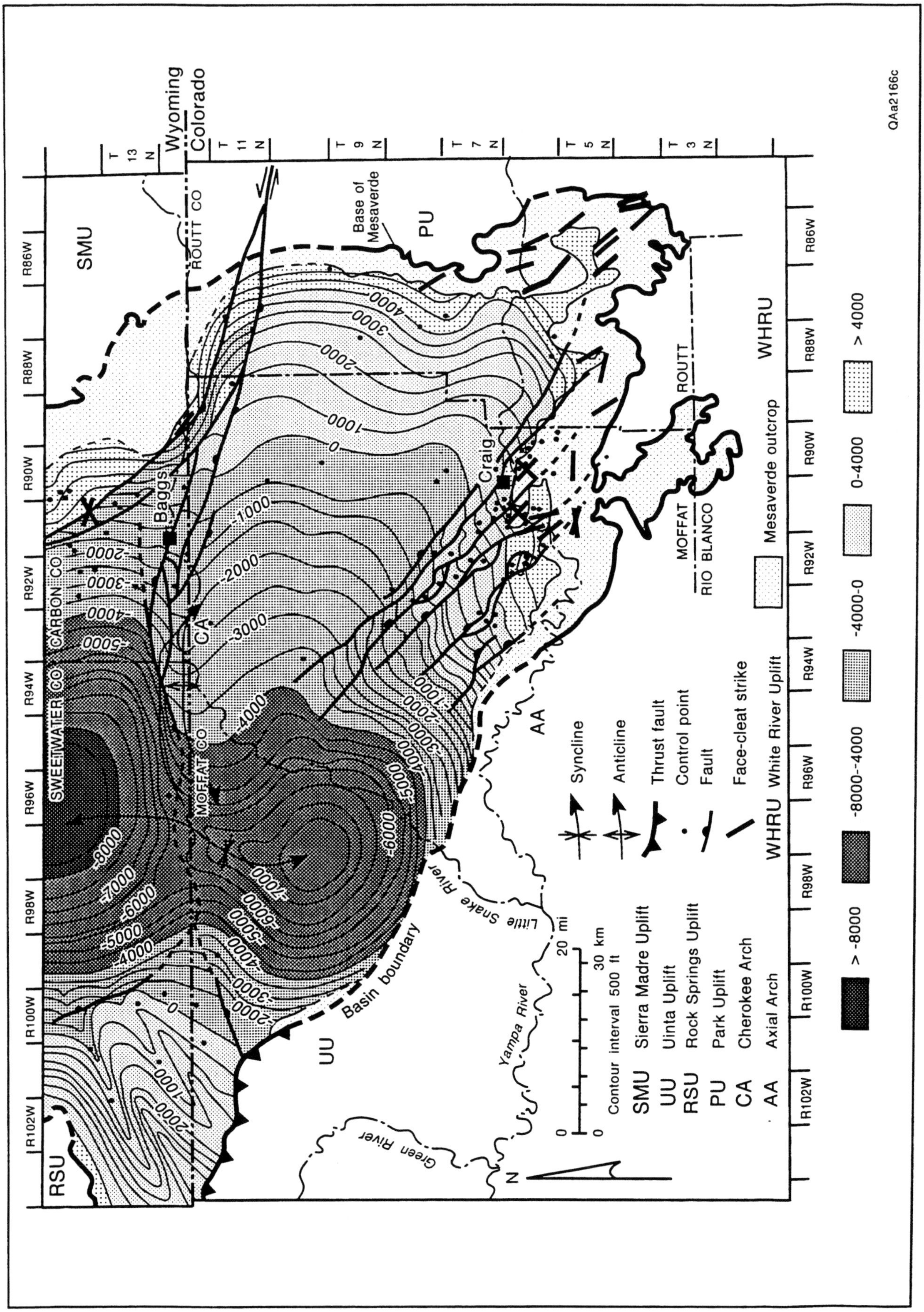

Figure 4. Structure map contoured on the base of the Williams Fork Formation, Mesaverde Group, Sand Wash Basin, showing face-cleat trends. Cleat observations are listed in table 2 or taken from Tyler and others (1991, 1992a), Khalsa and Ladwig (1981), and Boreck and others (1977).

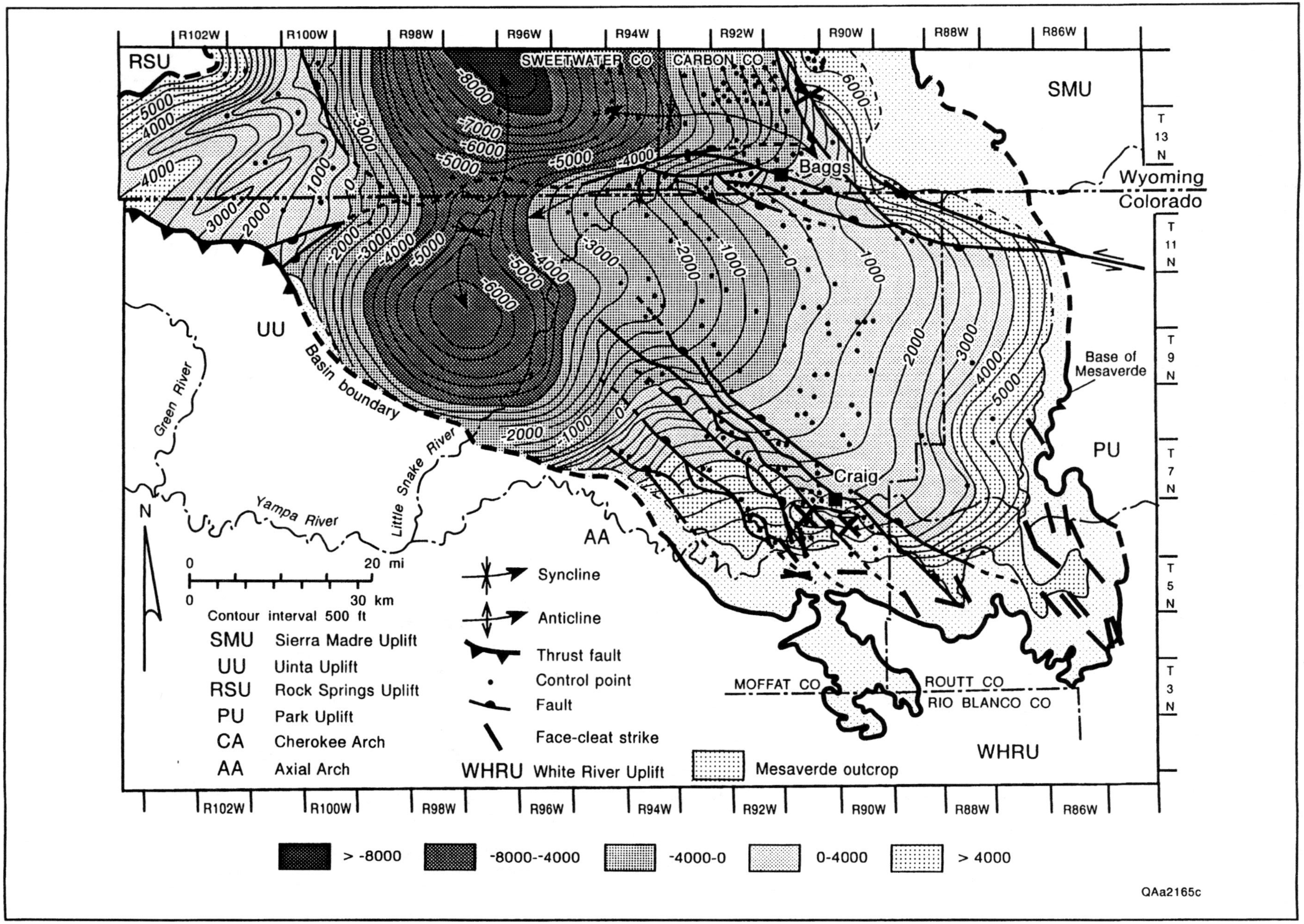

Figure 5. Structure map contoured on the top of the Williams Fork Formation, Mesaverde Group, Sand Wash Basin, showing face-cleat trends. Cleat observations are listed in table 2 or taken from Tyler and others (1991, 1992a), Khalsa and Ladwig (1981), and Boreck and others (1977).

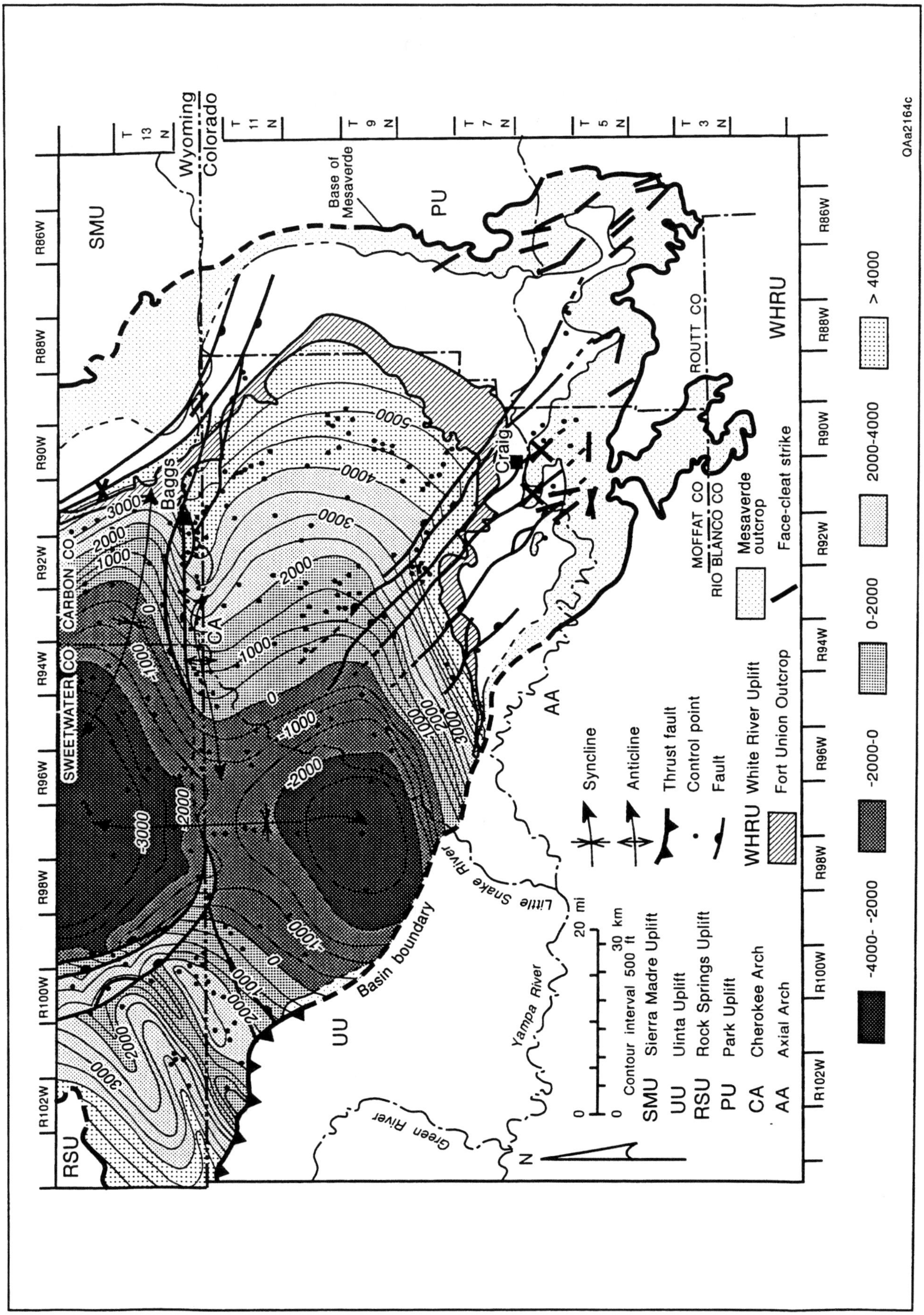

Figure 6. Structure map contoured on the base of the Fort Union Formation, Sand Wash Basin, showing cleat trends. Cleat observations are listed in table 2 or taken from Tyler and others (1991, 1992a), Khalsa and Ladwig (1981), and Boreck and others (1977).

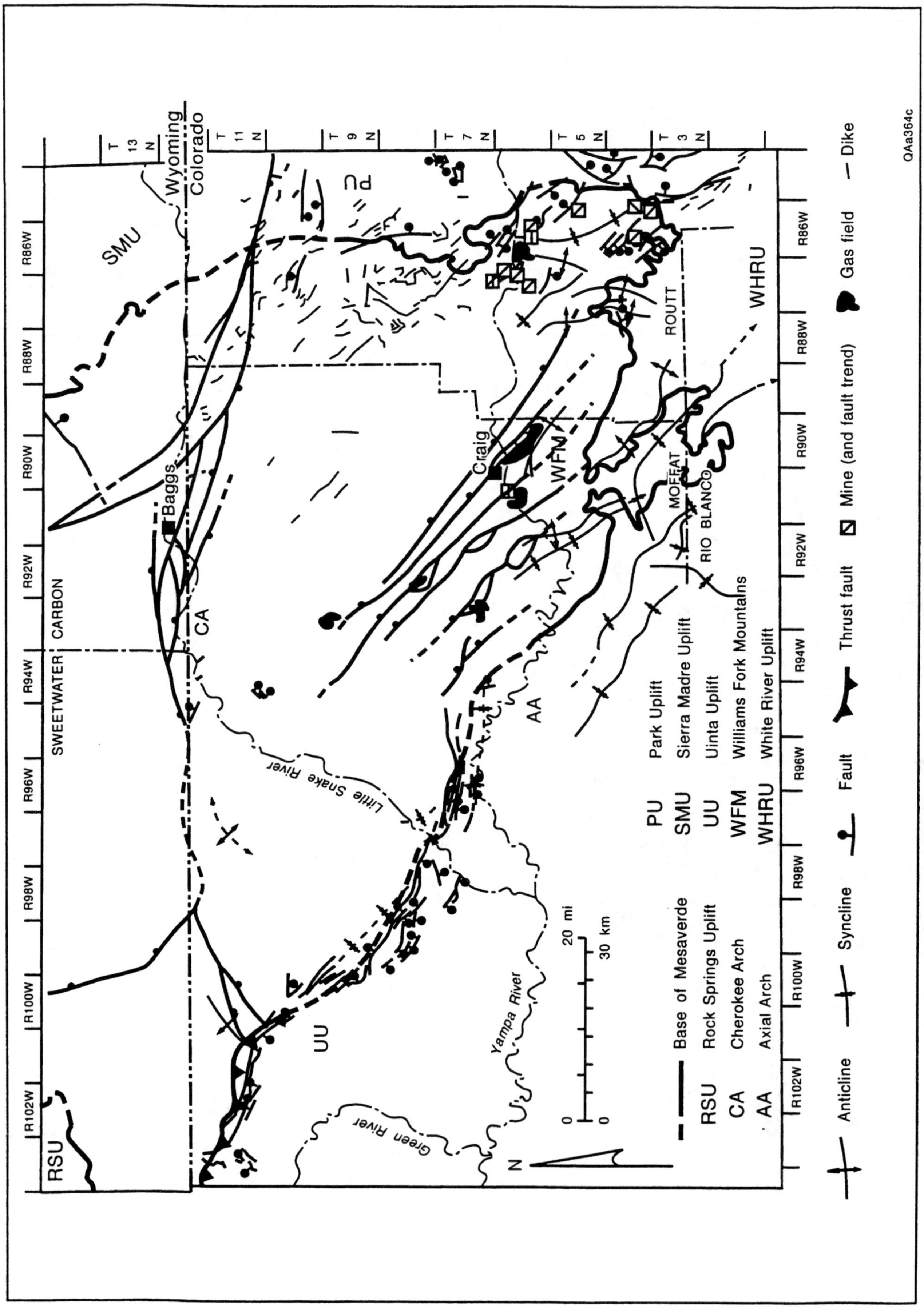

Figure 7. Map of tectonic elements in the Sand Wash Basin (modified from Hancock, 1925; Tweto, 1976a, 1979; Rowley and others, 1985). Locations of faults associated with coal mines and gas fields are shown. Note predominantly northwest trend of faults, folds, and dikes.

characterized by basin filling, normal faulting, and partial to complete collapse of basement uplifts, further modified the structural configuration of the basin (Hansen, 1965; Sales, 1983; Ryder, 1988).

Structural and Stratigraphic Settings of Coal-Bearing Formations

The coal- and coalbed methane-bearing formations in the Sand Wash Basin occur in Upper Cretaceous and lower Tertiary strata (Tyler and others, 1991, 1992a, b) (fig. 2). The Upper Cretaceous contains several coal-bearing, nonmarine stratigraphic units (Iles, Williams Fork, Almond, and Lance Formations) deposited in fluvial, delta-plain, and back-barrier settings, landward of delta-front and barrier-island systems (Haun, 1961; Asquith, 1970; Siepman, 1986; Roehler, 1990; Hamilton, this vol.). Structural maps contoured on the base and top of the Williams Fork Formation, the basin's major coal-bearing unit (figs. 4 and 5, respectively), show that the deepest portion of the Sand Wash Basin is on the flanks of the Uinta Uplift between T9N, R96W and T10N, R99W. The synclinal axis of the basin extends northward into the Washakie Basin where the base of the Williams Fork Formation is more than 17,000 ft (>5,182 m) deep. In the deepest part of the Sand Wash Basin, the base of the major coal-bearing Williams Fork Formation is about 7,000 to 7,500 ft (~2,134 to 2,286 m) below sea level (fig. 4). The base of the Mesaverde Group attains a maximum depth of about 13,500 to 14,000 ft (~4,115 to 4,267 m) below the surface. The top of the Mesaverde Group is about 11,500 to 12,000 ft (~3,505 to 3,658 m) below the surface. Coal-bearing strata are less than 5,000 ft (<1,524 m) deep on the flanks of the Rock Springs Uplift and on the western margin of the Sand Wash Basin; they crop out along the southern and eastern margins. The basin covers an area of approximately 5,600 mi^2 (~14,493 km^2) (Tyler and others, 1991) as defined by the outcrop trace of the base of the Mesaverde Group (figs. 4 and 5).

Lower Tertiary coal-bearing units include the Fort Union (Paleocene), Wasatch (Eocene), and Green River (Eocene) Formations (fig. 2). The major coal and coalbed methane targets in the Tertiary rocks of the Sand Wash Basin occur in the lower coal-bearing unit of the Fort Union Formation. In the deepest portion of the basin (T9N–T10N, R97W), the top of the Fort Union Formation is approximately 8,500 ft (~2,591 m) below the surface, with the base of the Fort Union Formation about 3,000 ft (~914 m) below sea level (fig. 6). The Fort Union Formation coal beds crop out along the flanks of the Rock Springs Uplift and on the eastern and southern margins of the Sand Wash Basin. Tertiary coal-bearing strata are buried at maximum depths ranging from 7,000 to 9,500 ft (2,134 to 2,896 m) in the Washakie Basin (McDonald, 1975).

Iles and Williams Fork Formations

The Iles Formation in the Sand Wash Basin consists of shelf and coal-bearing deltaic deposits (Boyles and Scott, 1981). The thickest seams (individual seams as much as 10 ft [3 m] thick) trend northeastward, parallel to the paleoshoreline. Thinner Iles coal beds, 3 to 6 ft (1 to 2 m) thick, overlie thin (<5 ft [<1.5 m] thick) crevasse splay sandstones that were local platforms for peat accumulation in interchannel areas.

The Williams Fork Formation consists of wave-dominated deltaic, back-barrier, and fluvial deposits (Boyles and Scott, 1981; Siepman, 1986; Hamilton, this vol.). The thick sandstone sequences served as platforms for peat accumulation (Siepman, 1986). Williams Fork coal beds occur in as many as 14 seams, with a net-coal thickness of as much as 220 ft (67 m) and maximum-coal thickness of 45 ft (14 m). Net-coal thickness trends are dominantly strike-elongate (northeast-oriented, parallel to the paleoshoreline), with minor dip-elongate (northwest-oriented) components (Siepman, 1986; Hamilton, this vol.). Williams Fork Formation coals are the Sand Wash Basin's prime coalbed methane target (Hamilton, this vol.).

Almond Formation

In outcrop along the Rock Springs Uplift and in subsurface studies, the Almond Formation ranges from 500 to 800 ft (152 to 244 m) in thickness. The Almond Formation contains as much as 35 ft (11 m) of coal; average coalbed thickness in the lower part of the Almond Formation is 8 to 12 ft (2.4 to 3.6 m) (Glass, 1981), whereas average coalbed thickness in the upper part of the Almond Formation is only 2 to 4 ft (0.6 to 1.2 m) (Roehler, 1988). East of the Rock Springs Uplift, the Almond Formation grades seaward into north-trending barrier-island sandstones (Weimer, 1965; Roehler, 1988, 1990). Coal beds have an average thickness of 3 ft (1 m) and are present at the top of at least four barrier-island sandstones. These coal beds split where they override tidal-inlet sandstones (Roehler, 1988). Upper Almond net-coal thickness ranges from 6 to 12 ft (1.8 to 3.6 m) in three to four seams, and these seams are potential coalbed methane targets.

Lance Formation

The Lance Formation, the youngest Cretaceous stratigraphic unit in the Sand Wash Basin, overlies and intertongues with nearshore-marine deposits of the Fox Hills Sandstone and consists of brackish and nonmarine shales, lenticular sandstones, and coal beds (Land, 1972). The Lance Formation is approximately 800 to 1,000 ft (~244 to 305 m) thick in the southern Sand Wash Basin, and about 200 ft (~61 m) thick in the northwest. Coal beds are thicker and more abundant in the lower part of the Lance Formation above the platform Fox Hills sandstone and range in thickness from a few inches to 20 ft (a few centimeters to 6 m). However, these coal beds have lateral extents limited to only a few hundred feet and are therefore unimportant as coalbed methane targets.

The Lance Formation is separated from the overlying Fort Union Formation by a massive fluvial sandstone sequence. The thick sandstone sequence, referred to herein as the *massive Cretaceous and Tertiary (K/T) sandstone unit,* contains the regional K/T unconformity, overlies and intertongues with the upper part of the Lance Formation, and underlies and intertongues with the lower coal-bearing unit of the Fort Union Formation. On geophysical logs the massive K/T sandstone unit is recognized by its blocky-log signature, thicknesses of hundreds of feet, and stratigraphic position below the coal-bearing Fort Union Formation. The massive K/T sandstone unit is correlatable throughout the basin and north into the Washakie Basin (Hettinger and others, 1991).

Fort Union Formation

In the Sand Wash Basin, the Fort Union Formation contains north- and northeast-trending, fluvial sandstones and floodplain coal beds. Net-coal thickness in the Fort Union Formation ranges from 0 to 90 ft (0 to 27 m) in as many as 12 seams at depths of as much as 8,000 ft (2,438 m) below the surface (Tyler, this vol.). Net-coal thickness and coal-seam continuity are greatest in the lower Fort Union Formation, where coals formed on floodplains above fluvial sandstones that served as platforms for peat accumulation. Coal beds are thicker and more numerous above these sandstones. Fort Union Formation coal beds are potential coalbed methane targets (Tyler, this vol.).

Wasatch Formation

The Wasatch Formation exhibits net-sandstone trends and depositional systems similar to those of the underlying Fort Union Formation (McDonald, 1975). The main body of the Wasatch Formation near the Rock Springs Uplift consists of 1,500 to 2,500 ft (457 to 762 m) of conglomeratic fan-delta deposits that grade eastward into fluvial sandstones, floodplain and lacustrine shales, and minor coal-bearing floodplain deposits (Roehler, 1965a; Sklenar and Anderson, 1985). Wasatch Formation coal beds are few and thin and are therefore minor coalbed methane targets.

Green River Formation

The Green River Formation (Eocene) is the youngest coal-bearing formation in the Sand Wash Basin. It intertongues with the underlying Wasatch Formation and consists of fluvial, paludal, floodplain, and lacustrine deposits. However, Green River lacustrine deposits are much more extensive than those in the Wasatch Formation (Surdam and Stanley, 1980). During deposition of the Green River Formation, a widespread lake system evolved in the basin; short-lived swamps are reflected by numerous, thin (<5 ft [<1.5 m] thick) and discontinuous coal beds grading laterally into carbonaceous shales. Coal beds of the Green River Formation are thin and discontinuous and are minor coalbed methane targets.

Faults, Folds, and Fracture Patterns

The subsurface and surface structures of the Sand Wash Basin are characterized by northwesterly and westerly striking faults of diverse origins, strong northwesterly striking anticlinal and synclinal folding, and a complex history of fracture genesis. Three major fault systems occur within the Sand Wash Basin, as mapped on the Williams Fork and Fort Union Formations. A west-east-trending fault system is associated with the Cherokee Arch to the west of Baggs; a north- and northwest-trending fault system is located to the east of Baggs; and a northwest-trending fault system occurs to the northwest and southeast of Craig (figs. 4 through 6). The orientation of fold axes generally parallel the major faults, showing a gradual shift from north-south on the eastern margin of the basin, to more northwest-southeast in the western and central parts of the Sand Wash Basin, suggesting shifting maximum horizontal stresses. Natural fractures (cleats) similarly record a complicated genetic history resulting from Laramide and post-Laramide structural deformation. These fault, fold, and fracture systems, and the thrusts and faults that bound the uplifts surrounding the Sand Wash Basin, result in a highly complex structural grain both within and along the margins of the Sand Wash Basin (fig. 7).

Faults and Folds

Faults in the Sand Wash Basin may contribute to coal permeability and conventional trapping of gas. Oil and gas fields occur on north-, northwest-, and west-trending faulted structures on the flanks of the Cherokee and Axial Arches and in the center of the basin. The west-east-trending Cherokee Arch, located to the north of the Wyoming–Colorado state line, is a westward-plunging anticline cut by numerous faults which are herein termed the *Cherokee Arch fault system* (figs. 4 through 7). Structural contours drawn on top of the Mesaverde Group and the Fort Union Formation reveal a major west-east-trending fault that splays out toward the west and east, producing a complex normal and reverse fault system, having a left-lateral strike-slip component. The fault system extends for at least 30 mi (48 km) in a west-east direction and is as much as 8 mi wide (13 km) between T12N, R96W and R90W (figs. 4 through 7). Downthrown blocks are generally on the northern side of the faults and total displacement across the system may be as much as 2,500 ft (762 m). A small horst with a throw of approximately 400 ft (122 m) occurs in T12N, R92W. Lateral shearing on strike-slip faults has also created local upthrust structures (Stone, 1975), resulting in a complicated thrust, reverse, and normal fault system (figs. 4 through 7).

To the east and northeast of the Cherokee Arch fault system, two major northwesterly trending faults, herein termed the *Savery fault system* (Scott and Kaiser, this vol., fig. 43), extend for approximately 40 mi (~64 km) along the margins of the Mesaverde Group outcrop. Maximum displacement across the fault system may be as much as 2,500 ft (762 m); downthrown blocks are on the western side of the faults. The easterly trending Cherokee Arch fault system and the northwesterly trending Savery fault system, when traced to the southeast, connect with a strike-slip fault system that crops out within the Sierra Madre Uplift (Petroleum Information Corporation, 1992). Seismic lines were not available for in-depth structural studies of these fault systems. Cronoble (1969) proposed that the high-angle to vertical faulting occurred after the deposition of the Fort Union and Wasatch Formations, making the fault systems active during or after early Eocene.

The southwestern part of the basin is bordered by thrust, reverse, and strike-slip fault systems that extend approximately 80 to 100 mi (~129 to 161 km) and parallel faults on the northeast flank of the Uinta Mountains and Axial Arch (fig. 7). Northwest of Craig (figs. 4 through 8), a major system of faults, herein termed the *Cedar Mountain fault system*, has been recognized in the subsurface from geophysical logs and seismic lines provided by Union Pacific Resources. The fault system is at least 10 mi (16 km) wide and extends approximately 30 mi (~48 km) northwest and 15 mi (24 km) southeast of Craig. The projection of the fault system boundaries southeastward coincides with thrust and reverse faults mapped from seismic data (Livesey, 1985, and herein), prominent northwest-trending lineaments, and northwest-trending outcrop segments of the Mesaverde Group–Lewis Shale contact (fig. 5). As many as six faults trend northwest, range in length from 5 to 45 mi (8 to 72 km), are all nearly vertically downthrown to the northeast, parallel one another, and individually have throws between 500 and 1,800 ft (152 and 549 m) for a total displacement across the system of more than 5,000 ft (1,524 m) on the top of the Mesaverde Group (fig. 8). In the subsurface, left-lateral strike-slip motion is also inferred on the system's largest fault in T7N, R94W from Williams Fork Formation cross-fault sedimentology and stratigraphy. The Williams Fork Formation in the USA 15-1 well in the upthrown block is more akin to the fluvial-dominated Williams Fork to the northwest than to coastal plain Williams Fork typical of the Craig area and found immediately across the fault in the downthrown block. In outcrop in the Cedar Mountain area, the faults are mapped as normal faults, where Miocene strata (Browns Park Formation) are downfaulted against Paleocene strata (Fort Union Formation) (Tweto, 1979).

Large predominantly northwest- and north-trending folds occur along the southeast border of the basin (Tweto, 1976). These folds include the northwest-trending Williams Fork, Beaver Creek, Breeze, and Buck Peak anticlines in the west (Hancock, 1925) and the more northerly trending Tow Creek, Oak Creek, Fish Creek, and Sage Creek anticlines on the far eastern margin of the Sand Wash Basin (Bass and others, 1955). Northwest faults, 5 to 10 mi (8 to 16 km) long, are recorded parallel to the fold axes on surface geologic maps (Bass and others, 1955; Hancock, 1925; Tweto, 1976). Smaller faults, oblique to the folds, have also been reported. Faults with displacements of 2 to 215 ft (0.6 to 66 m) have been mapped in the subcrop in 11 abandoned and 6 operating mines (table 1). The majority of these in-mine faults trend northwest although minor east, west, and northeast faults (and a few northwest-trending dikes) have also been mapped. Some of the fault displacements observed in coal mines may be the result of strike-slip motion as indicated by slickensides (Robson and Stewart, 1990). In addition, faulting has also created fracture swarms within or between several fault planes that parallel the fault traces. Northwest-trending faults also appear on subsurface maps of gas fields such as Buck Peak, Craig Dome, Great Divide, Tow Creek, and Big Gulch.

Preexisting structural grain (zones of weakness), dating back to the Precambrian, may have controlled the orientation of the fault and fold systems. During the Laramide Orogeny there was extensive thrust, reverse, normal, and/or strike-slip faulting. Maximum horizontal stresses were oriented either southwest-northeast or northwest-southeast, or both. Uplift occurred again

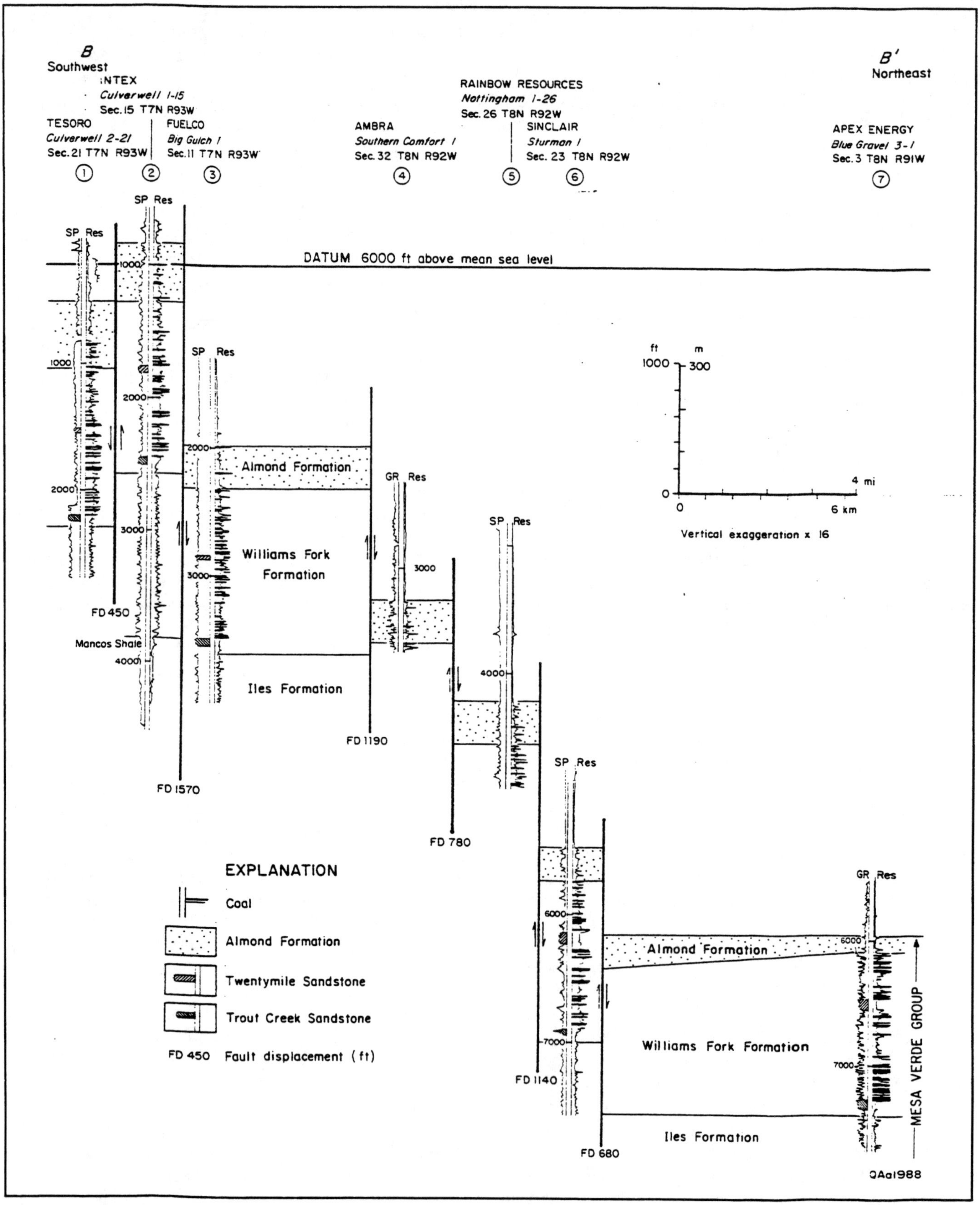

Figure 8. Structural cross section through the Cedar Mountain fault system, drawn on the Mesaverde Group, Sand Wash Basin. Fault blocks downthrown to northeast. Throws range from 500 to 1,500 ft (152 to 457 m). Total displacement across the system is about 5,000 ft (~1,524 m).

Table 1. Coal mine faults in the Sand Wash Basin.

MINE NAME	SECTION	TWP	RGE	MINE TYPE	MINE STATUS	FAULTS MAPPED	FAULTS TRENDS	FAULT THROWS (ft)	NOTES
Apex	21,22	4N	86W	U	Abd.	2	NW,NW	25 and 100	
Bear River	11,2	6N	87W	U	Abd.	4	EW,NW,NW,NW	One at 8	
Blair	SW,NW 10	6N	91W	U	Abd.	1	ENE		
Curtis	NE,SW 22	6N	86W	U	Abd.	1	NW	70	
Denton Strip	20,21	6N	86W	S	Abd.	3	NNE,WNW,WNW		
Hammond	SE,NW 34	7N	87W	U	Abd.	2	NNE,NW		
Harris	16,21,28, 15,22,27	6N	87W	U	Abd.	15	NW	2–215	
Keystone	19	4N	85W	U	Abd.	3	SW,WNW,NW		
	24	4N	86W						
Lenox	SW 22	6N	86W	U	Abd.	1	N7 W	6	
Pinnacle	35,36	4N	86W	U	Abd.	4	WNW,NNW,NNW,NW	3–20	NW-trending dike
	1,2	3N	86W						
Wadge	9,10,15	6N	87W	U	Abd.	10	NW & 9 at NNW	4–7	9-ft-wide NNW dike, coked coals 18–22 ft per side
Seneca Strip	2,3,10,11	6N	87W	S	Act.	4	EW,NNW	EW is 40–60 NNW are 4	
Energy No. 1	13	5N	86W	S	Act.	1 major	NW-SE	60–100	
Edna	19,30,31	5N	85W	S	Act.	many	major faults NW		Pyrite in Wolf Creek and Wadge seams
	36	5N	86W				smaller faults NE		
	7,18,19	4N	85W						
Trapper	5 & 6	5N	90 W	S	Act.	1	EW		
	1,2,3,4,5	5N	91W						
	30–32	6N	90 W						
Eagle Mine	31,32	6N	91W	U	Act.	7	WNW	10–40	
	5,6	5N	91W						
Foidel Creek	32	5N	86W	U	Act.	1	NW	6	

S Surface mine
U Underground mine
Act. Active mine
Abd. Abandoned mine

during the Oligocene, and following the Laramide Orogeny in Miocene time, a tensional stress regime was present in the Sand Wash Basin, and extensional deformation occurred. Extensional faulting continued at a diminishing rate into Quaternary time.

Fracture Patterns

Permeability in coal is largely due to the occurrence of fractures (cleats) and faults. Cleat and fault characteristics were recorded in the Sand Wash Basin from field observations in the Mesaverde Group and Fort Union Formation coal beds (at approximately 26 stations, principally in the southeast corner of the basin), literature, and core descriptions (Colorado Oil & Gas Commission's well files). Additional information on faults was obtained from maps (including 100 maps of abandoned coal mines) and mine permits.

Cleat Types

According to the definition of Tremain and others (1991a, b), the first formed and commonly better developed fracture set in coal is the *face cleat*; generally, face cleats are the more prominent because their fracture traces are long and have smooth, planar surfaces. The less well developed, more irregularly shaped set, which abuts the face cleat, is the *butt cleat*. Observations in the Sand Wash Basin commonly show well-developed face cleats; butt cleats are less pronounced. The face and butt cleats are usually mutually perpendicular. They are also generally perpendicular to the coal bedding planes, although some cleat inclinations may vary between 60° to 90°. In addition to the face and butt cleats, occasionally crosscutting third- and fourth-order cleats were observed. Also, striated and sheared coals were seen at several locations, as were curved cleats and conchoidal fractures (table 2).

Face-Cleat Strikes from Outcrop Observations

Boreck and others (1977) measured north to northwest face-cleat directions in seven mines in the southeast part of the basin. They reported face-cleat striking at 003° at the Apex Mine (T4N, R86W), 353° at the Edna Strip (T4N, R85W), between 300° and 335° at four Energy Strip pits (T5N, R86W–R87W), and 315° at the Seneca Strip (T6N, R87W). Khalsa and Ladwig (1981)

Table 2. Cleat and fracture observations in the Sand Wash Basin.

AREA		LOCATION	LITHOLOGY	CLEAT/ FRACTURE	FM	FACE CLEAT ORIENTATION	COMMENTS
Haybro roadcut	SW	Sec. 17, T4N, R85W	Coal	Face cleat	Ki	350	3rd, curved, fracture swarms
Abandoned mine		Sec. 18, T4N, R85W	Sandstone	Joint	Ki	85	
Edna mine	SW	Sec. 24, T4N, R86W	Coal	Face cleat	Kwf	320	
Edna mine	SWSW	Sec. 31, T5N, R85W	Coal	Face cleat	Kwf	315	
Foidel Creek roadcut	NW	Sec. 28, T5N, R86W	Coal	Face cleat	Kwf	87	
Foidel Creek	NE	Sec. 29, T5N, R86W	Sandstone	Joint	Kwf	310	
Seneca Wolf Creek Seam	SENW	Sec. 1, T5N, R87W	Coal	Face cleat	Kwf	319	
Seneca Wadge Seam	SESE	Sec. 1, T5N, R87W	Coal	Face cleat	Kwf	316	
Hayden Gulch	SE	Sec. 17, T5N, R88W	Coal	Face cleat	Kwf	330	Bright, friable coal
Hayden Gulch	NE	Sec. 30, T5N, R88W	Coal	Face cleat	Kwf	290	
Berry Gulch	NW	Sec. 28, T5N, R89W	Coal	Face cleat	Kwf	335	
Jeffway Gulch	SE	Sec. 9, T5N, R90W	Coal	Face cleat	Ki	89	Dull coal
Jeffway Gulch	NE	Sec. 9, T5N, R90W	Coal	Face cleat	Kwf	95	Moderately dull coal
Jeffway Gulch	NE	Sec. 9, T5N, R90W	Coal	Face cleat	Kwf	90	Boney coal
Roadcut Highway 13	NE	Sec. 17, T5N, R91W	Coal	Face cleat	Ki	90	
Roadcut past McGregor	SW	Sec. 9, T6N, R86W	Coal	Face cleat	Kwf	350	3rd, 4th, and sheared & conchoidal
Roadcut past McGregor	SW	Sec. 9, T6N, R86W	Coal	Butt cleat	Kwf	82	
Meadow No.1 mine roadcut	SW	Sec. 12, T6N, R87W	Coal	Face cleat	Ki	338	
Mt.Harris roadcut	SE	Sec. 15, T6N, R87W	Coal	Butt cleat	Kwf	50	3rd & 4th cleats
Mt.Harris roadcut	SE	Sec. 15, T6N, R87W	Coal	Face cleat	Kwf	330	
Walker mine	SE	Sec. 17, T6N, R90W	Coal	Longer cleat	Kwf	44	2 mutually abutting & crosscutting cleats
Walker mine	SE	Sec. 17, T6N, R90W	Coal	More frequent cleat	Kwf	316	
Yampa River	NE	Sec. 16, T6N, R91W	Coal	More frequent cleat	Kwf	320	2 mutually abutting & crosscutting cleats
Yampa River	NE	Sec. 16, T6N, R91W	Coal	Longer cleat	Kwf	45	
Roadcut Eagle mine	NE	Sec. 31, T6N, R91W	Coal	Face cleat	Kwf	325	
Roadcut Eagle mine	NE	Sec. 31, T6N, R91W	Coal	Butt cleat	Kwf	50	
Williams Fork	SESE	Sec. 33, T6N, R91W	Coal	Face cleat	Kwf	327	
Hancock location 552-553	SWSW	Sec. 23, T6N, R92W	Coal	Face cleat	Kwf	323	Good, finely cleated coal
Hancock location 131-133	NW	Sec. 17, T6N, R93W	Coal	Face cleat	Kwf	318	Weathered coal
Routt Cty. 52 roadcut	NE	Sec. 27, T7N, R87W	Coal	Face cleat	Ki	320	Dull coal, some curved cleats, ss dikes
Franz mine	W 1/2	Sec. 36, T8N, R87W	Sandstone	Joint	Ki	326	
Seymour Mine	SW	Sec. 17, T8N, R89E	Coal	Face cleat	Kfu	330	Thick, bright, well-cleated coal
Thomas mine, Savery WY	SE	Sec. 5, T12N, R89W	Coal	Butt cleat	Kmv	63	Bright coal, slickensides, calcite & pyrite
Thomas mine, Savery WY	SE	Sec. 5, T12N, R89W	Coal	Face cleat	Kmv	312	Good cleat, some curved, fault zones in butt direction

Kfu Fort Union Formation
Kwf Williams Fork Formation
Ki Iles Formation
Kmv Mesaverde Group

also measured northwest face-cleat strikes of 300°–312° at the Denton Strip (T6N, R86W) and 314°–320° at the Eagle No. 5 underground mine (T6N, R89W) (figs. 4 through 6). Face-cleat orientations measured at 26 stations in the Sand Wash Basin (table 2, fig. 6) generally trend northwest (Tyler and others, 1991, 1992a, b, c; Laubach and others, 1992a, b, c), parallel to the current maximum horizontal stress direction (Zoback and Zoback, 1989) and the major northwest-trending faults in the area (fig. 7). However, the northwest strike of the face cleats shifts south of Craig. On the Yampa River, on Highway 789, and at the abandoned Walker Mine (table 2), mutually crosscutting and abutting cleats strike northwest and northeast. We tentatively interpret these cleats as two major, possibly contemporaneous face-cleat sets (Laubach and others, 1992a, b, c) that are related to the shifting of the stress regime during Cenozoic times. These mutually abutting, crisscrossing fracture sets may also enhance permeability (Tremain and others, 1991a, b). To the south in T5N, R90W–R91W, on Highway 13, face-cleat strike is nearly east-west (fig. 7); major faults south of Craig, in T4N, R91W–R92W, also strike east-west.

Cleat Spacing and Fracture Swarms

In many coals, cleat spacing varies with coal rank, coal lithotype, ash content, and bed thickness (Ammosov and Eremin, 1960) and with position relative to structural deformation. The spacing between cleats is currently used in reservoir modeling as an indicator of potential fracture permeability (Mavor and others, 1991), although fracture interconnectedness and tortuosity are more important controls. Interconnectedness and tortuosity, however, cannot be measured in core, but outcrop characterization can facilitate prediction of cleat attributes and coal permeability in fractured reservoirs.

To standardize cleat-spacing description, Tremain and others (1991a, b) divided cleats into four groups based on their relationship to coal lithotypes or bedding surfaces. (1) Master cleats cut through an entire coal seam including thin, noncoal interbeds, and spacing can be highly variable. (2) Primary cleats are contained within, but extend the entire height of a coal lithotype. Since they are large, master and primary cleats may be significant for fluid migration, but they are only rarely seen in core because of their wide spacing. (3) Secondary cleats are more frequent than primary cleats, but they do not cut an entire lithotype. (4) Tertiary cleats are very closely spaced fractures that occur between secondary cleats, generally with heights of less than 0.5 inch (<1.27 cm).

Master and primary cleat spacing in high-volatile C bituminous Mesaverde coals studied in mine and outcrop are highly variable. Master cleat spacing in the Mesaverde group coals at the Seneca Mine varied from 6 to 12 ft apart. Primary spacing between face cleats in Mesaverde coals at the Edna and Energy surface mines is 2.4 to 6 inches (6.1 to 15 cm) (Boreck and others, 1977). Primary spacing between face cleats at the Haybro roadcut is 0.5 to 1 inch (1.27 to 2.54 cm); at Hayden Gulch, 1 inch (2.54 cm); and at the Thomas Mine, 12 inches (30.5 cm) (see table 2 for locations). Spacing between secondary cleats in high-volatile C bituminous Mesaverde coal outcrops is generally between 0.25 and 0.5 inch (0.6 and 1.27 cm). One- to 2-inch (2.54- to 5.1-cm) cleat spacing was recorded in a Mesaverde coal at 4,914 to 4,923 ft (1,498 to 1,500 m) in the Helmerich and Payne Colorado State No. 1-31 well (Sec. 31, T7N, R88W). Spacing between butt cleats in a Fort Union coal, from approximately 5,000 ft deep (~1,524 m) in the Chevron Federal Land Bank (F.L.B.) No. 15-4C, is 0.25 inch (0.6 cm). Thin vitrain bands in Fort Union coals, as in most coals, are closely cleated, on the order of less than 0.25 inch (<0.6 cm) in a Fort Union coal from 2,072 to 2,077 ft (631 to 633 m) in the F.L.B. No. 1-29 well (Sec. 29, T7N, R92W).

An intensification of cleat frequency and interconnectedness was observed in fracture swarms and fault zones parallel to the butt-cleat direction at the Haybro roadcut and Thomas Mine. Some mine operators have reported an influx of methane associated with fracture swarms, and high gas contents have been measured that are associated with faults along the southern part of the basin (Kaiser and others, this vol.). Northwest-striking face cleats, faults, and folds parallel the current maximum horizontal stress direction and are permeable pathways that probably contributed to the migration and conventional trapping of gas. In addition, areas with a pronounced northwest structural grain might be areas of high-permeability anisotropy where in-mine horizontal wells perpendicular to the face-cleat direction or refracturing of coal seams (in which new fractures may initiate perpendicular to the original fracture [Palmer, 1993]) may be effective completion methods. On the other hand, where mutually abutting face cleats occur, as observed at the Walker Mine and Yampa River sites (see table 2) or where northeast-trending (butt-cleat direction) fracture swarms, faults, and folds such as those noted on the eastern side of the basin create low cleat permeability anisotropy, cavity completions might be favored.

Cleat Mineralization

Minerals deposited in cleats can obstruct the permeability of fracture systems in coal seams. Although cleats in many Sand Wash Basin coals lack cleat-filling minerals in outcrop, several instances of mineralization have been noted (table 3). Calcite fills some cleats at the Thomas Mine (table 3) near Savery, Wyoming. Accompanied by pyrite, calcite lines cleats in a few coals cored in the USGS C-IC-H well (Sec 23, T4N, R93W). Calcite was also reported *throughout cleats* in an 8-ft (2.4-m) coal cored in the Helmerich & Payne Colorado State No. 1-31 well (table 3). Hancock (1925) reported several instances of selenite (gypsum) along joint planes in blocky coals at a few old mines and prospects (table 3). Minor amounts of pyrite are also frequently reported in coal mines and cores. The pyrite occurs as isolated rosettes on cleat surfaces in fresh coal samples. Reddish-brown staining in outcropping coals and associated sandstones may be weathered pyrite formerly present in the cleats and joints.

Stress Regime

The interpretation and timing of the orientation of the principal shortening direction in the Sand Wash Basin are controversial. The major compressive force during the Laramide Orogeny was either east-west (Livesey, 1985), southwest-northeast (Gries, 1983), west-southwest–east-northeast (Stone, 1975), or northwest-southeast (Laubach and others, 1992a, b, c). Dynamic analysis of subsurface and surface structures in northwestern Colorado (Stone, 1975, and herein) indicates that the structural patterns of the Sand Wash Basin are consistent with the regional tectonic patterns of the Rocky Mountain foreland. The complex fault, fold, and fracture patterns were either produced by east-northeast–west-southwest and/or northwest-southeast maximum horizontal compressional stresses and lateral shearing at depth, during the early Laramide Orogeny. Spatially, the orientation of the faults and fold axes shows a gradual change from almost north-south on the eastern margin of the Sand Wash Basin, adjacent to the Sierra Madre–Park Uplift, to a more northwest-southeast orientation in the western and central parts of the basin, suggesting a counterclockwise shifting of the maximum horizontal stresses about a vertical axis from east-west to northeast-southwest. This shift in maximum horizontal stresses can be interpreted as progressing to a nearly north-south-reoriented maximum horizontal stress that could have resulted in the emplacement of the west-east-oriented, northerly thrusted Uinta Uplift (Stone, 1975).

Laramide and post-Laramide stresses associated with the genesis of natural fractures (cleats) in the Sand Wash Basin have similarly shifted about a vertical axis both spatially and with time. Upper Cretaceous and Early Tertiary coal beds are cut by a complex network of extensional fractures and cleats. Fracture data reveal at least three principal face-cleat strikes and corresponding stress orientations in the Mesaverde Group coals of the Sand Wash Basin. The first formed and better developed fracture set (face cleat) in coal beds of the Iles Formation commonly strike north in T4N, R85W (table 2), northwest in T6N–T8N, R87W (table 2), and west in T5N, R90W (table 2). The dominant face cleats in the Williams Fork Formation strike north to northwest in T4N–T6N, R86W (table 2), northwest to west-northwest in T5N, R87W–R89W (table 2), and west-northwest to west in T5N–T6N, R90W–R92W (table 2). The youngest and less well developed butt cleats generally strike northeast. Regionally the Iles Formation has dominant face-cleat strikes of west and north-northwest, and the Williams Fork Formation has a dominant northwest face-cleat strike with evidence for mutually abutting northeast face-cleat strikes. A gradual change in face-cleat strike from northwest to more north on the eastern edge of the basin suggests a shifting of the principal horizontal stresses through Cenozoic time. A record of Laramide and post-Laramide stress rotation has also been documented for joints in the Piceance and Washakie Basins (Verbeek and Grout, 1986; Grout and Verbeek, 1992a, b).

In summary, southwest-northeast and/or northwest-southeast maximum horizontal stresses have produced the northwest structural grain to the Sand Wash Basin and resulted in northwest-striking thrust, normal, and strike-slip fault systems and fold axes. This maximum horizontal stress is consistent with stresses that produced the regional tectonic patterns of northwestern Colorado; this stress also produced the northwest-striking face cleats, which parallel the fault systems and fold axes of the Sand Wash Basin.

The present stress regime of the Sand Wash Basin is extensional and lies within the Cordilleran stress province of Zoback and Zoback (1989) between the Colorado Plateau interior and the southern Great Plains stress province (fig. 9). Using sparse stress measurements, Zoback and Zoback (1989) suggest that the maximum horizontal compressive stress orientation is north-northwest in the Sand Wash Basin.

Conclusions

1. Major coal-bearing and coalbed methane targets occur in Upper Cretaceous strata of the Williams Fork Formation (Mesaverde Group) and lower Tertiary strata of the lower coal-bearing unit, Fort Union Formation.

2. The complex subsurface and surface structures of the Sand Wash Basin are characterized by important northwest- and west-striking faults of diverse origins, strong northwest-striking anticlinal and synclinal folding, and a complex history of fracture genesis.

3. Northeast-southwest and/or northwest-southeast maximum horizontal stresses are expressed dynamically in a combined thrust and strike-slip (wrench) fault segmentation of the Sand Wash Basin accompanied by

Table 3. Cleat mineralization in the Sand Wash Basin.

AREA	LOCATION	FM	COAL DEPTH INTERVAL	MINERALS	REFERENCE
USGS C-IC-H coal core hole	Sec. 23, T4N, R91W	Kwf	176–800 ft	P,C,R	Tremain and Toomey, 1983
Prospect, Locn. No. 251	Sec. 29, T4N, R92W	Kwf	Surface	G	Hancock, 1925
Battle Era Mine, Locn. No. 47	Sec. 14, T4N, R94W	Kwf	Surface	G,P	Hancock, 1925
Prospect, Locn. No. 405	Sec. 6, T5N, R92W	Kwf	Surface	G	Hancock, 1925
Helmerich & Payne State 1-31	Sec. 31, T7N, R88W	Kwf	4,914–4,923 ft	C	COGCC Files
Energy Reserves Van Doren No. 1	Sec. 29, T7N, R90W	Kwf	4,649–4,706 ft	P,R	Tremain and Toomey, 1983
Thomas Mine	Sec. 5, T12N, R89W	Kmv	Surface	C,P	Personal observation
Meridian No. 11-23 State	Sec. 23, T12N, R92W	Kfu	1,530–1,790 ft	P	COGCC Files
Mountain Fuel No. B-6 Allen	Sec. 33, T12N, R97W	Kfu	5,420–5,890 ft	P	COGCC Files
Edna Mine	Sec. 36, T5N, R86W	Kwf	Surface	P	COMLRD Files
Energy Strip No. 1A	Sec. 32, T5N, R86W	Kwf	Surface	P	Boreck and others, 1977

COGCC Colorado Oil and Gas Conservation Commission
COMLRD Colorado Mined Land Reclamation Division
G Gypsum
P Pyrite
R Resin
C Calcite

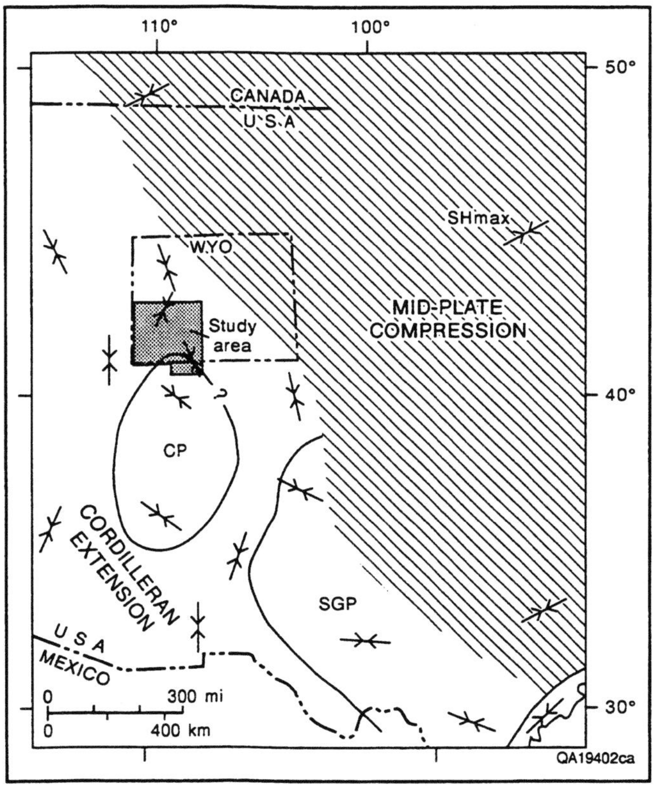

Figure 9. Stress province map showing major stress province boundaries in the vicinity of the Sand Wash Basin. Inward-pointing arrows indicate SHmax direction. CP = Colorado Plateau stress province; SGP = Southern Great Plains stress province. Study area is near the boundary between the Cordilleran extensional province and the midplate compressional province. Modified from Zoback and Zoback (1989).

intense anticlinal folding of Late Cretaceous and earlier sediments, resulting in significant structural relief.

4. High water production (Kaiser and others, this vol.) from coalbed methane wells in the Sand Wash Basin indicates high permeability. This permeability may in part reflect open northwest-trending face cleats in the southeast part of the basin, where face cleats are parallel to current maximum horizontal stress directions.

5. Local areas of crosscutting and mutually abutting face cleats and fracture swarms in the butt-cleat direction may be areas of increased cleat connectedness and decreased permeability anisotropy. Such areas could be favorable targets for completion techniques such as cavity completions that have proved successful in the northern San Juan Basin. Fracture swarms and faults could also create conventional traps for gas.

6. Laramide and post-Laramide stresses shifted about a vertical axis both spatially and with time from east-northeast, northeast, and north to northwest.

7. Prediction of fracture patterns at depth in the Sand Wash Basin may be improved by additional detailed outcrop characterization of fracture attributes. Core holes do not contain oriented core, and very few wells are available for core analysis.

Upper Cretaceous Mesaverde Group

Stratigraphy and Coal Occurrence of the Upper Cretaceous Mesaverde Group, Sand Wash Basin

Douglas S. Hamilton

Abstract

The upper Mesaverde Group is divided into the Williams Fork and Almond Formations. The Williams Fork Formation is the most important coal-bearing unit in the Sand Wash Basin. It is divided into four genetic depositional sequences, each bounded by regionally extensive, low-resistivity shale markers. Units 1 through 3 were characterized by linear shoreline systems in the easternmost part of the basin that were bounded landward by coastal plain systems, which in turn graded landward into fluvial systems. Unit 4 deposition was dominated by a mixed-load fluvial system. The Almond Formation is a minor coal-bearing unit that was deposited as a wave-dominated delta system.

The thickest, most laterally extensive coals occur in Williams Fork Units 1 and 2, the two lowermost genetic units. These coals are concentrated in the eastern half of the basin, east of the Little Snake River, and are thickest near Craig, where net coal thickness of Unit 1 averages 90 ft (27.4 m) and Unit 2 averages 40 ft (12.2 m). Average net coal thicknesses of Units 3 and 4 are 30 and 40 ft (9.1 and 12.2 m), respectively, but the coals are less continuous. Unit 3 and 4 coals are thickest northwest of Craig, and Unit 4 contains the only appreciable coal west of the Little Snake River. Variability in coal continuity was demonstrated by the coal-seam profiles. Whereas some seams could be traced by their characteristic density and gamma-ray log profiles over most of the eastern half of the basin, others could be correlated only when grouped as broad coal packages. Unit 1 and 2 coals are continuous from the subsurface to the outcrop belts in the south and northeast and are thus potential conduits for basinward flow of ground water. Unit 3 and 4 coals are less continuous in the subsurface and are unlikely to provide potential for interconnected aquifer systems. Data are scarce on Almond coal distribution, but three areas—(1) west of Craig, (2) southeast of the Rock Springs Uplift, and (3) west of the Sweetwater-Carbon County line—contain net coal thickness of as much as 25 ft (7.6 m).

Ideal conditions for peat accumulation and preservation occurred on the coastal plain of Units 1 and 2 immediately landward of equivalent shoreline sandstones. Bypassing coarse clastic sediment, maintenance of high water-table levels, and optimum subsidence combined in this setting. Gradual westward thinning of Unit 1 and 2 coals toward the coastal-plain/alluvial-plain transition is explained by a lowering water table associated with the rise in surface gradient of the alluvial piedmont. Coals also thin to the east as they overrode the shoreline sandstones. Marine conditions ultimately limit coal distribution to the east. Unit 3 coals, despite occupying a similar coastal plain setting, are not as thick or extensive as those of Units 1 and 2 probably because the area of sediment bypass was smaller and subsidence rates were not optimal. Unit 4 coal distribution was controlled by a mixed-load fluvial system, and peats accumulated in isolated interchannel areas between fluvial axes. Almond coals are located behind shoreline sandstones and between dip-oriented distributary facies.

Introduction

A general assessment of all coal-bearing intervals of the Sand Wash Basin was undertaken to target those units with greatest potential for coal-bed methane production. The Williams Fork Formation was quickly identified as containing the thickest, most extensive, and greatest number of coal seams, and was selected as the principal focus of the study. The Almond Formation was not studied in detail.

The Williams Fork and Almond Formations form the upper part of the Mesaverde Group, which is a major pre-Laramide, Upper Cretaceous coal-bearing sequence (fig. 10). During the Upper Cretaceous, the area of the Sand Wash Basin was occupied by the Western Interior Seaway, which received clastic sediment in cycles initiated by tectonic uplift and loading of the Overthrust Belt to the west. Sedimentation patterns are also thought to be influenced by eustatic sea-level fluctuations (Kauffmann, 1977).

***In* Kaiser, W. R., and others, 1993, Geologic and hydrologic controls on coalbed methane: Sand Wash Basin: The University of Texas at Austin, Bureau of Economic Geology, topical report prepared for the Gas Research Institute under contract no. 5091-214-2261 (GRI–92/0420), p. 23–49.**

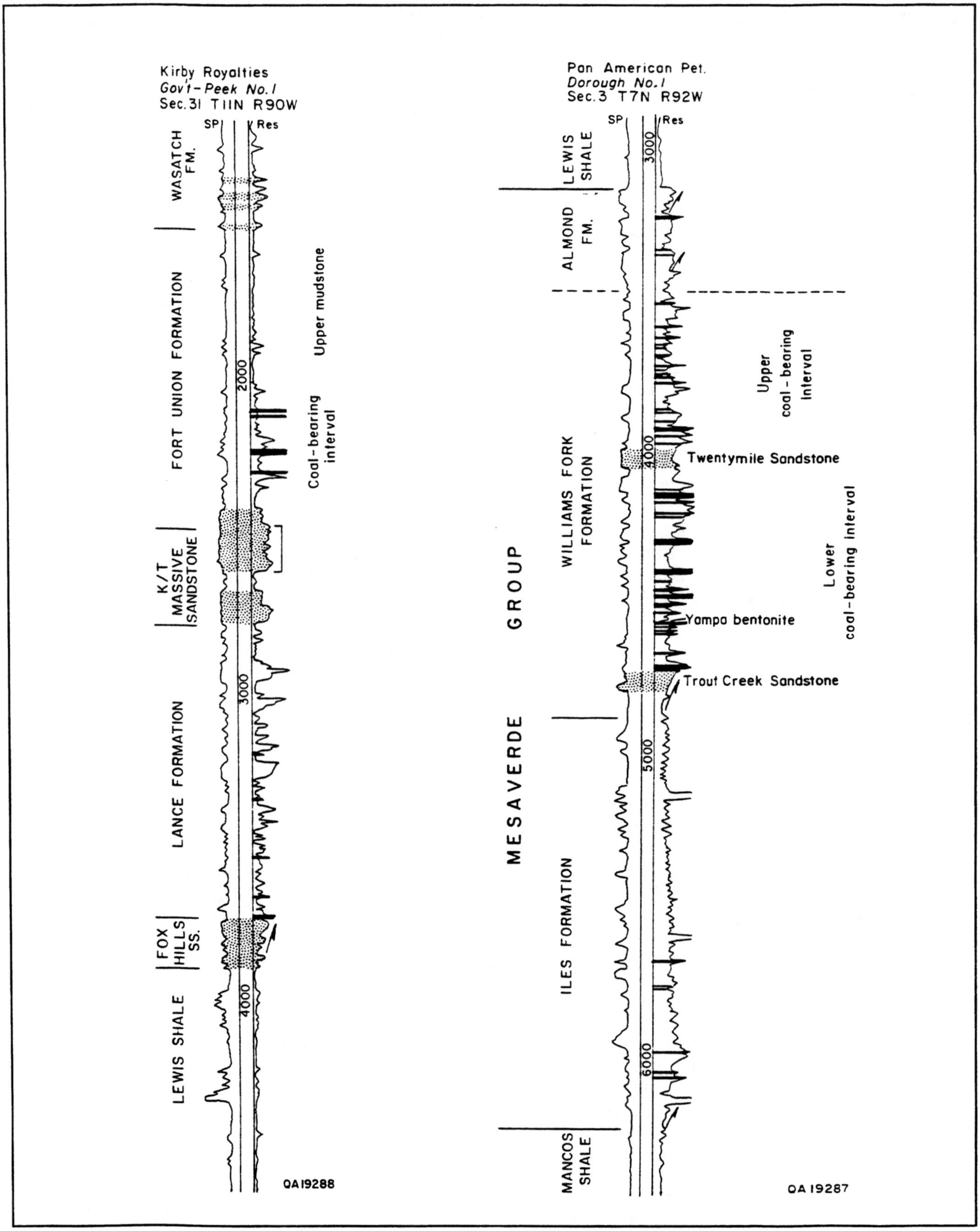

Figure 10. Stratigraphy of the Sand Wash Basin. The most important coal-bearing interval is the lower Williams Fork Formation. Fort Union coals are also important.

The first step taken in this study was to establish a stratigraphic framework in which detailed and meaningful analysis of the coals, and their enclosing sediments, could be carried out. A genetic approach to stratigraphic analysis was applied to the upper Mesaverde Group. The genetic stratigraphic framework then provided the basis for delineation of the major depositional systems and mapping the distribution and thickness of the coals. This stratigraphic framework further provided a basis for investigating the depositional controls on coal occurrence.

Genetic Approach to Stratigraphic Analysis

The best way to achieve meaningful understanding of a sedimentary sequence is to identify and investigate strata that are genetically linked. Ideally, genetic units to be mapped should be correlatable over widespread areas and should have been deposited during discrete episodes of general tectonic, climatic and/or base level stability (Galloway, 1989). Such units are the fundamental time-stratigraphic increments of the basin fill, and they provide the foundation for establishing a correlation framework and construction of basic lithofacies maps necessary for further interpretation. More detailed analysis allows the delineation of the component depositional systems, which are characterized by specific geometries and bedding architecture (Galloway and Hobday, 1983) that are readily determined from subsurface data.

Depositional systems are also characterized by specific processes of sediment dispersal that can be observed directly in modern-day analogs. Herein lies the real strength of the genetic approach. Recognition of the depositional system, in conjunction with an understanding of its sediment dispersal processes, provides a powerful guide for predicting lateral changes in geometry and distribution of the framework sandstone facies and associated coal-bearing mud rocks. Detailed understanding at the facies level is the ultimate objective in coalbed methane research because it is at this scale that (1) the lateral continuity and thickness of the coalbed reservoirs are determined, and (2) the basin's fluid migration pathways, including the target coalbed gases and the produced waters, are established.

Interrelated with the task of delineating the major genetic units is recognizing the hiatal surfaces that bound these units. The hiatal surfaces record major interruptions in basin depositional history and represent significant periods of nondeposition or very slow clastic accumulation. The bounding surfaces are generally easily recognized in marginal marine basin settings where widespread marine shales separate successive progradational clastic wedges (Frazier, 1974; Galloway, 1989). However, recognition of the bounding surfaces in nonmarine basin-fills is more problematic, and possibly only the erosional unconformities provide obvious sequence boundaries. More subtle, conformable bounding surfaces are important but require considerably more intensive investigation for recognition.

Recognition of the principal bounding surfaces of the upper Mesaverde genetic sequences was relatively straightforward in the eastern half of the basin. The basin occupied a marginal marine setting along the western edge of the Western Interior Seaway during upper Mesaverde deposition, and the successive clastic wedges are bracketed by transgressive marine flooding surfaces. Defining bounding surfaces in the continental facies to the west was more difficult but still possible.

Genetic Stratigraphy of the Upper Mesaverde Group

A genetic stratigraphic framework was established for the upper Mesaverde Group. The unit can be divided genetically into the Williams Fork and Almond Formations. The Williams Fork Formation can be further subdivided into four genetic units, Units 1 through 4 (fig. 11), each representing a discrete depositional episode within the basin's history. The genetic units are bounded by regionally extensive, low-resistivity shale markers that have been mapped from the southeastern margin of the basin to at least as far west as T13N, R102W on the southern flank of the Rock Springs Uplift, and to the north beyond the limits of the study area (figs. 12 and 13). The shale markers are attributed to marine flooding surfaces in the basinward direction (east and southeast), where they are easily recognized separating aggradational coal-bearing coastal plain facies of one depositional episode from overlying upward-coarsening progradational sequences of the next. In the landward direction, genesis of the shale markers is less clear. Either the marine flooding events that punctuated the Williams Fork extended further west than is generally recognized, or the controls on the flooding events, such as shutting-off sediment supply, similarly affected the nonmarine environment and are also recorded by low-resistivity shale markers indicative of sediment starvation. Units 1 through 4 are thus true genetic depositional sequences as defined by Galloway (1989) because they are depositional units bounded by flooding surfaces (and their nonmarine correlative surfaces).

Comparison with Traditional Stratigraphy

The Williams Fork Formation as defined here varies from the traditional stratigraphy in three main ways.

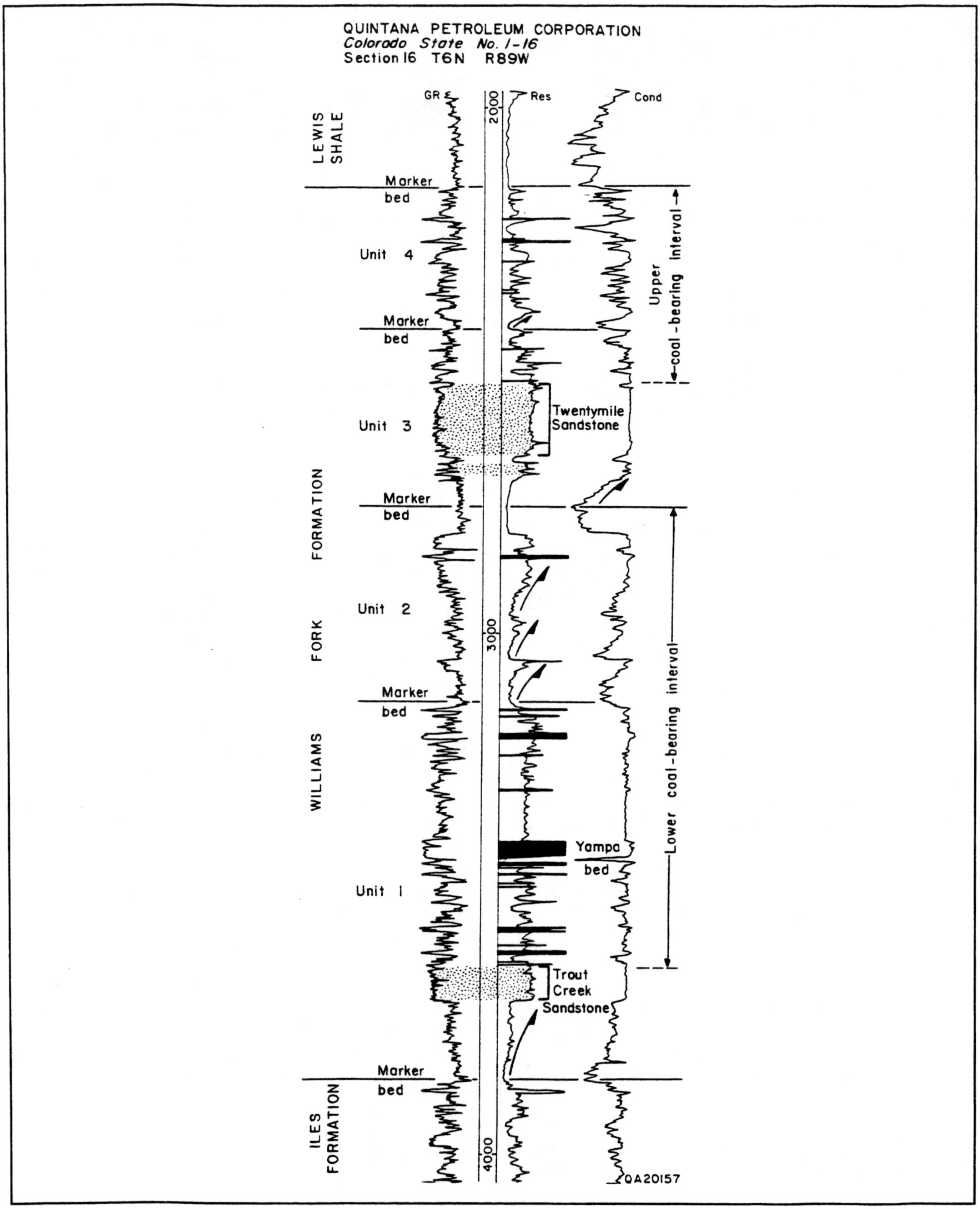

Figure 11. Genetic stratigraphy of the upper Mesaverde Group in the eastern Sand Wash Basin. Coal beds are identified on an accompanying density log. Surfaces bounding genetic units are defined by regionally extensive, low-resistivity shale marker beds.

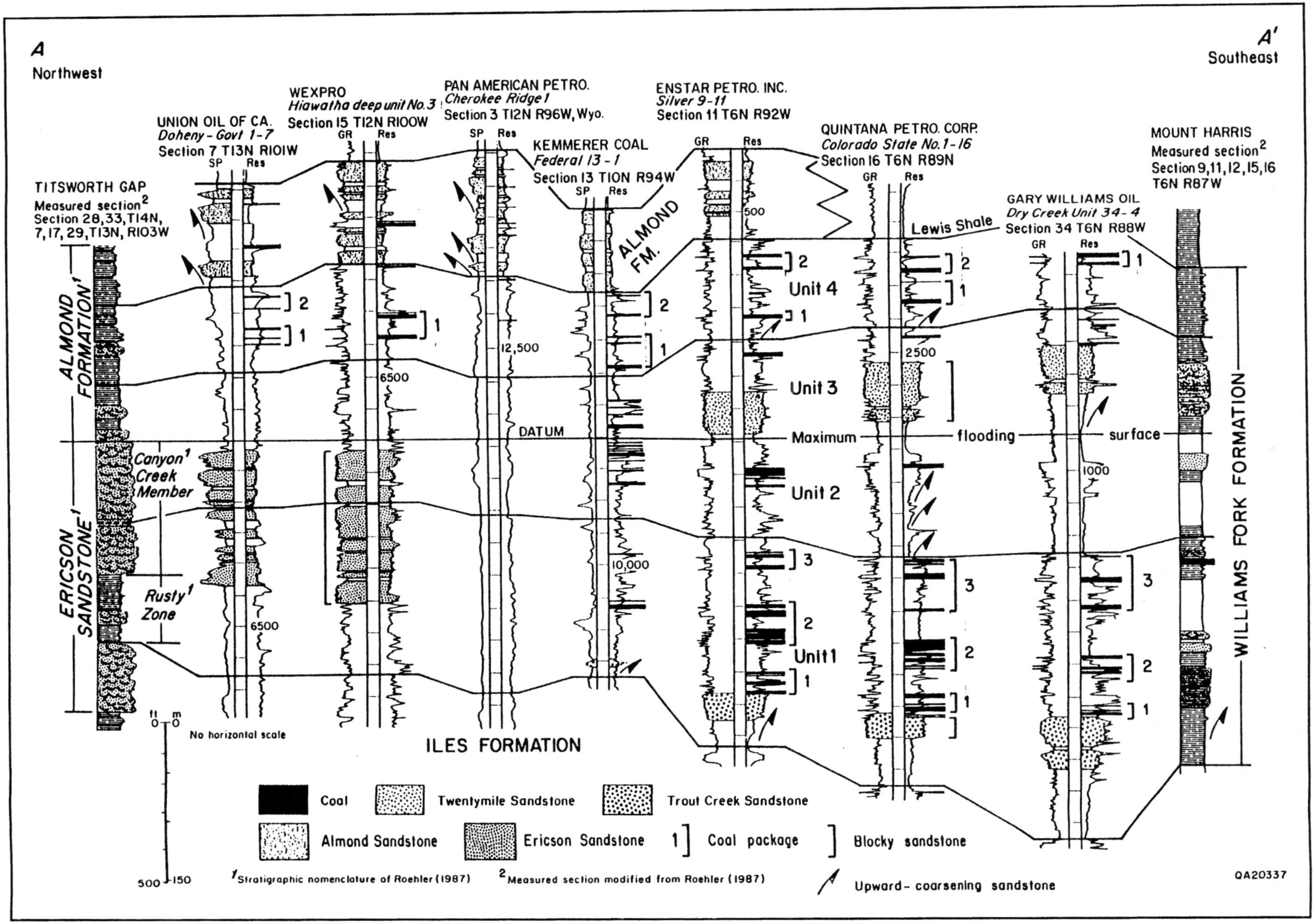

Figure 12. Northwest-southeast cross section of the upper Mesaverde Group, Sand Wash Basin, illustrating genetic stratigraphy of the Williams Fork Formation and coal occurrence. Unit 1 and 2 coals are lost westward of the transition to alluvial Ericson Sandstone at the basin's structural center.

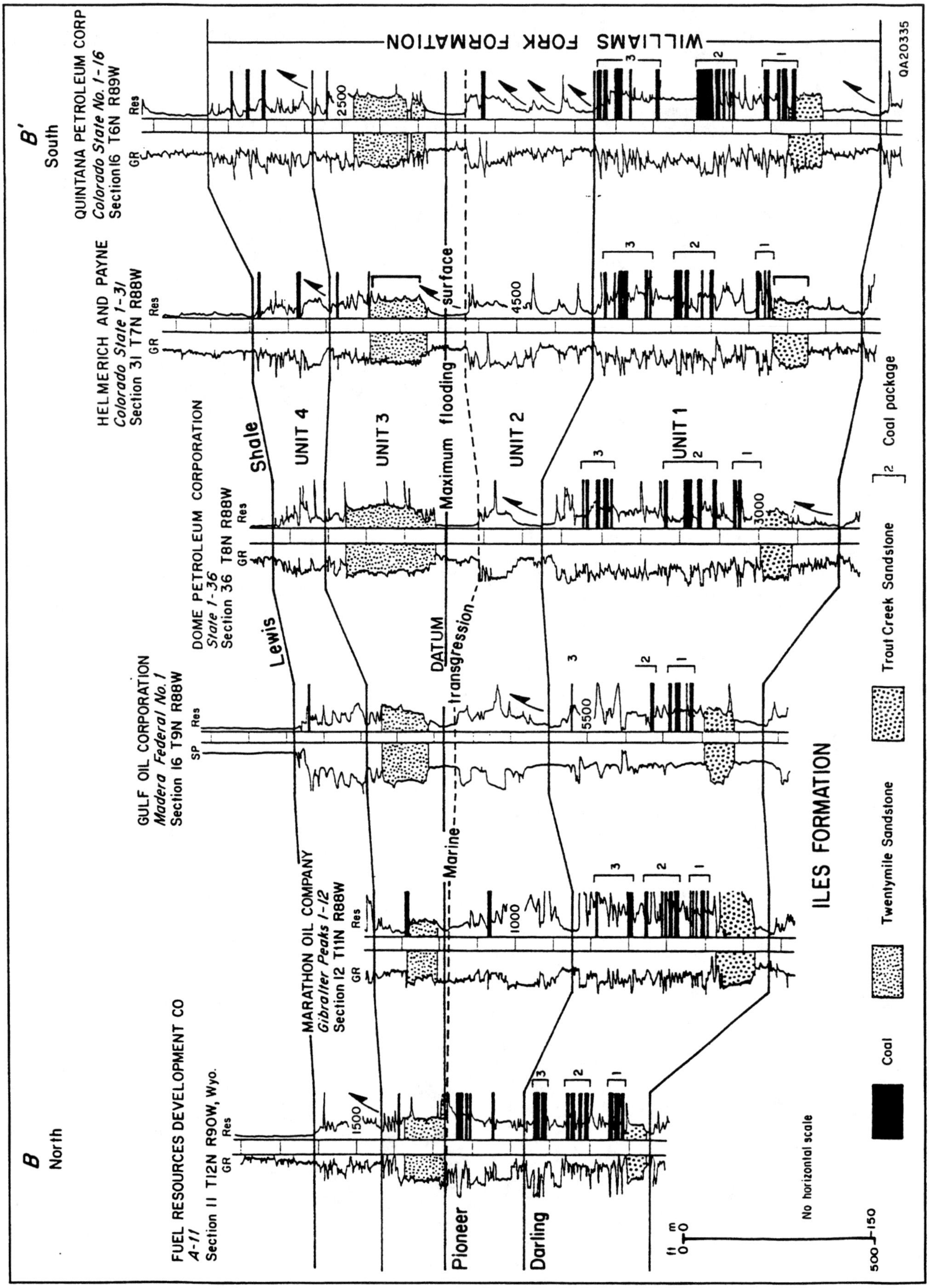

Figure 13. North-south cross section of the upper Mesaverde Group, eastern Sand Wash Basin. The Williams Fork Formation extends northward into Wyoming.

1. The Trout Creek Shale and overlying Sandstone Member, which are traditionally assigned to the uppermost part of the underlying Iles Formation (Siepman, 1986) are in this study included with the Williams Fork Formation. Depositionally, the Trout Creek Shale/Sandstone couplet records an episode of marine transgression and subsequent progradation. Thus, the progradational Trout Creek sequence belongs genetically with the Williams Fork Formation (figs. 1 and 2).

2. The Williams Fork Formation is distinct or separated from the Almond Formation. In his published cross section, Roehler (1987) showed the Almond Formation as partially equivalent to the upper part of the Williams Fork Formation. The Almond Formation, as traditionally defined, includes two dissimilar sedimentary sequences, that is, a prominent aggradational sequence of interbedded sandstones, siltstones and coals, and an overlying, strongly progradational sequence of upward-coarsening and blocky sandstones with coal beds. Here we restrict the term Almond Formation to the upper, strongly progradational sequence and the Williams Fork to the underlying aggradational coal-bearing sequence (fig. 3). A regionally extensive, low-resistivity shale marker separates these two sequences, and the change in their character is evident on gamma-ray, spontaneous potential (SP), and resistivity logs. Genetically, the Almond Formation represents a barrier-bar/strandplain complex that lies above the main Williams Fork coal-bearing interval.

3. The genetic depositional sequences of the Williams Fork Formation (Units 1, 2, 3, and 4) cut across many of the traditionally defined lithological members. For example, the top of Unit 1 cuts through the middle of the Canyon Creek Member (fig. 3), and the top of Unit 2 cuts through the middle of the Pine Ridge Sandstone Member (as illustrated in Roehler and Hansen, 1989).

Coal Occurrence of the Upper Mesaverde Group

Coal Identification

Coals were identified in this study from geophysical well logs by low bulk density, low natural gamma response, very high resistivity, high neutron and density porosities, low sonic velocity, and/or low neutron count. Some combinations of these criteria were used because no uniform well log suite was available. Bulk density or sonic logs were run in most wells, and these are the most reliable logs for coal identification. However, natural gamma response was consistently low for all coal beds and was used in conjunction with very high resistivity, and shalelike SP response to operationally define coal in some wells.

Coal Seam Continuity

Continuity of the Williams Fork coals is variable. Some individual seams were correlatable in the subsurface throughout the eastern half of the Sand Wash Basin and extend to the southern and northeastern outcrop belts. Other seams could only be correlated extensively when grouped as broad coal packages. Data were too scarce to demonstrate continuity of Almond coals. Understanding coal seam continuity is critical to coal gas production and water production because (1) coal seams with considerable continuity provide pathways for diffusion and long-distance migration of coal gases and (2) continuous coals act as major aquifers.

Coal seams are correlatable because of their unique seam signature. They are biochemical sediments composed of discrete bands (or lithotypes) that are a function of the original peat-forming plants and the physical and chemical conditions that prevailed in the peatswamp. Coal-seam correlation is achieved by recognizing the unique seam signature in adjacent wellbores.

Seam signatures of some typical, laterally continuous, Williams Fork coals are illustrated in figure 14. The seam signatures are defined by the gamma-ray and density logs, which are sensitive to minor fluctuations in the coal-seam lithotypes. The seam 1 gamma-ray and density-log profile has a serrate key-like shape. Seam 2 is characterized by several splits that display an upward decrease in density. Seam 3 is recognized by its three parts, or plies, and the middle plie is consistently the most prominent. The top coal, seam 4, is characterized by its blocky signature. A number of discontinuous coals are also illustrated in figure 5. These show a featureless spike on both the gamma-ray and density logs.

Detailed discussion of individual Williams Fork coals will be framed within the context of their encompassing depositional system, which controls the coal distribution and thickness. Discussion of Almond coals is much less detailed.

Williams Fork Genetic Depositional Sequences

Unit 1

The lowermost genetic depositional sequence of the Williams Fork Formation, Unit 1, is a clastic wedge that

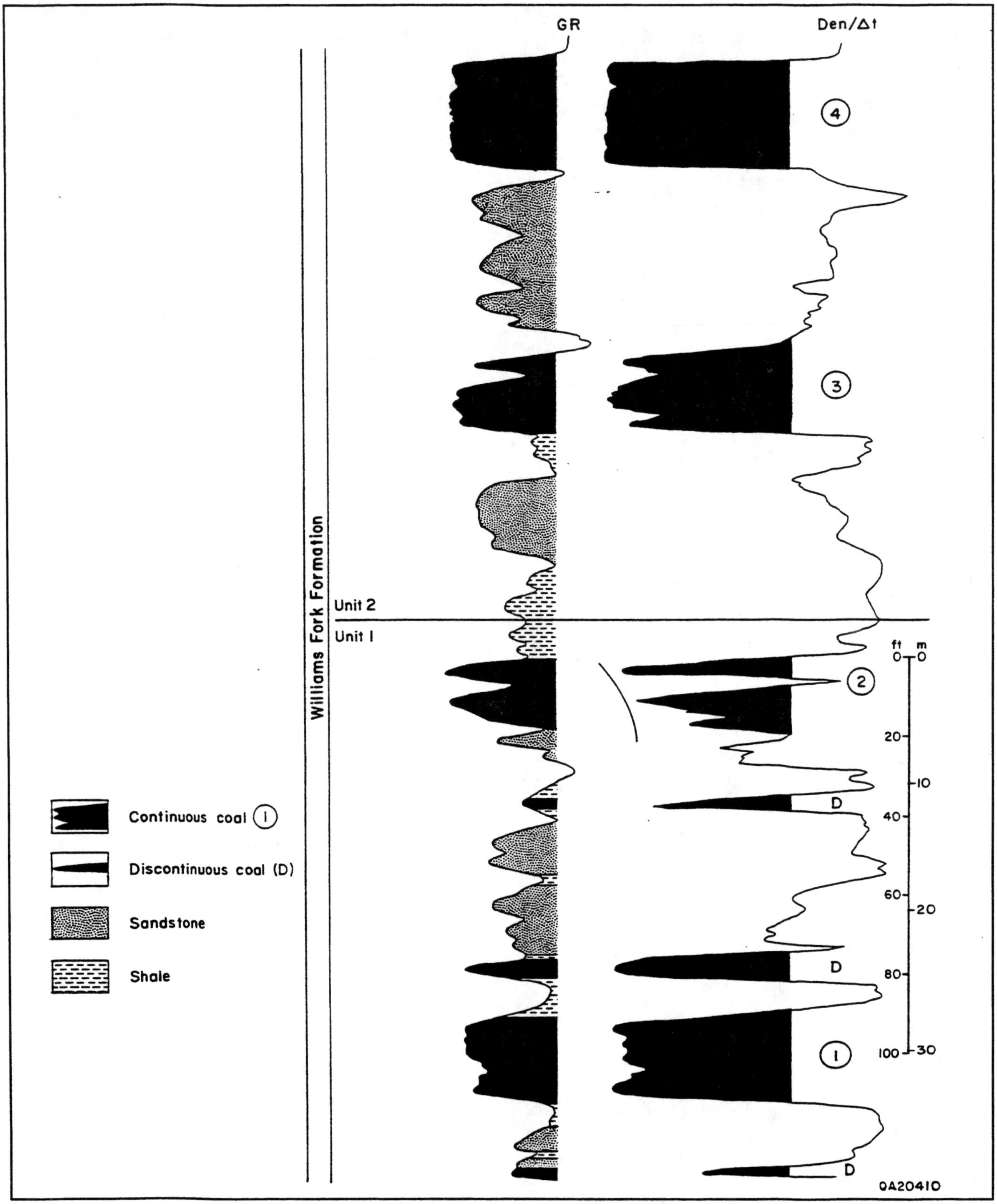

Figure 14. Density profile of typical Williams Fork coals. Coal seams can have a unique profile, or signature, owing to variations in their lithotypes. The lithotypes are a function of the original peat-forming plants and the physical and chemical conditions of the peatswamp. Gamma-ray and density logs are sensitive to minor fluctuations in coal seam lithotypes and thus provide a tool for correlation.

extended coal-bearing coastal-plain deposits to beyond the present-day basin margin. The unit is bounded by regionally extensive, low-resistivity shale markers. The lower bounding surface occurs near the base of the Trout Creek Shale Member in the eastern and southeastern parts of the basin, where the sequence is characterized by the upward-coarsening, progradational Trout Creek Sandstone Member and overlying aggradational coal-bearing rocks. There is a prominent facies change to the west as the coal-bearing strata are replaced by thick, stacked sandstone units and interbedded mudstones of the Ericson Sandstone (fig. 15). Stratigraphically, Unit 1 is equivalent to the Trout Creek Shale and Sandstone Members and lower one-third of the Williams Fork Formation in the eastern part of the basin, and the middle part of the Ericson Sandstone in the west. To the north, this unit is equivalent to the upper part of the Allen Ridge Formation (Roehler, 1987). Unit 1 thickness ranges from 900 ft (274 m) in the southeast, where basin subsidence was at a maximum, to 400 ft (122 m) in the northeast. Basin subsidence trends have a pronounced northeast-southwest alignment.

Depositional Systems

Three major depositional systems are recognized in Unit 1 from the geometry of framework sandstones and log facies mapping. A linear shoreline system dominates the easternmost part of the basin and is backed landward by a coastal plain system that grades westward into a mixed-load to bed-load fluvial system (fig. 15).

A number of parallel strike-oriented (northeast-southwest) sandstone-rich trends are apparent in the easterly shoreline system (fig. 15). This, coupled with the strong upward-coarsening log motifs, provides evidence of shoreline progradation. The shoreline system is backed by a sand-poor area (net sandstone less than 125 ft [38 m]) that defines the coastal plain system. The coastal plain was largely an area of sediment bypass, and the aggradational log patterns that characterize this system reflect thick coals and interbedded mudrocks. A dip-oriented sandstone-rich trend extending southeasterly from Baggs cuts across the coastal plain (fig. 15), and is interpreted as a distributary channel complex that fed sediment to the shoreline system. Log patterns of this zone are blocky and upward-fining, consistent with such an interpretation. The coastal plain passes landward (westerly) into the alluvial plain where contributary patterns in sandstone distribution define a major fluvial system (Ericson Sandstone). Log patterns are aggradational and associated with thick, stacked channel sandstones with interbedded floodplain muds.

Coal Stratigraphy

Unit 1 coals are the thickest and most extensive in the Sand Wash Basin. Three discrete coal packages are recognized, and each extends over the entire eastern part of the basin (fig. 12). The first (or lowermost) package immediately overlies, and is genetically related to, the Trout Creek Sandstone. Three coal seams from 3 to 10 ft (0.9 to 3 m) thick are typically present in this package, but as many as five much thinner (2 to 5 ft [0.6 to 1.5 m] thick) seams may be present locally, where seam splitting occurs.

The second coal package overlies the first and consists of two coal seams that can be correlated individually over most of the eastern part of the basin. Correlation is achieved by matching their characteristic profiles, as displayed on the gamma and density logs. Correlatability is further enhanced by the presence of the distinctive Yampa bentonite bed (fig. 11), which occurs within the lower of the two seams. The seams merge in T6N, R89W, where the combined coal thickness is 43 ft (13 m), but elsewhere seam splitting is common. Individual seam splits range from 5 to 25 ft (1.5 to 7.6 m) thick.

The third (or uppermost) coal package consists of as many as five seams ranging in thickness from 2 to 20 ft (0.6 to 6 m) (fig. 12). Correlation of individual seams in this coal package was possible only over an area of approximately 80 mi^2 (207 × 10^6 m^2) (T7-8N; R92W), where one 15 to 20 ft (4.5 to 6 m) seam had a characteristic gamma-ray and density-log profile.

Coal Distribution

Net coal thickness is at a maximum in the Craig area, where it is as much as 129 ft (39.3 m) thick and averages 90 ft (27.4 m) thick (fig. 16). Net thickness decreases westward and is absent along a line from T8N R95W to Baggs, Wyoming, approximately paralleling the course of the Little Snake River in Colorado. There is no significant Unit 1 coal west of the Little Snake River in the structurally deepest and most thermally mature part of the basin. Thinning also occurs in the southeasternmost part of the basin, where net coal thickness is 30 to 40 ft (9.1 to 12.2 m). The pronounced northeast-southwest alignment of coal-seam thickness trends parallels the basin subsidence trends. Unit 1 isopachs indicate a northeast-southwest depositional strike and gradual thickening of the section to the southeast. Coals are also thin along a narrow, dip-oriented zone extending in a southeasterly direction through Baggs, where they are partially replaced by stacked sandstone units. Although some of the coals are replaced, the net coal thickness map indicates that

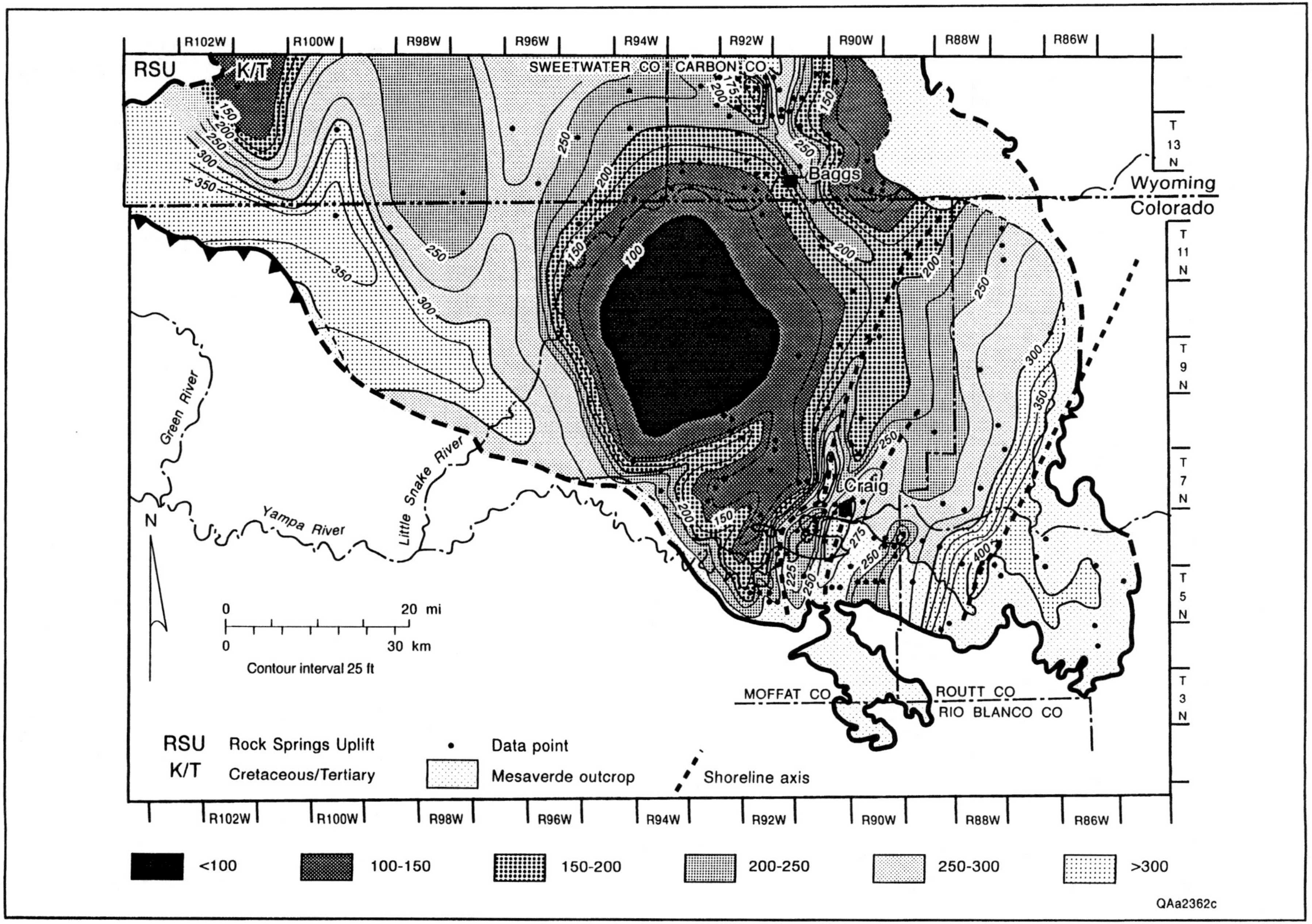

Figure 15. Net-sandstone map of Unit 1, Williams Fork Formation. A number of strike-oriented (northeast-southwest) linear clastic shorelines are evident, backed landward by a coastal plain system (net sandstone less than 125 ft [<38 m]), which grades westward into a large fluvial system (Ericson Sandstone). A dip-oriented distributary channel system extends southeastward from Baggs.

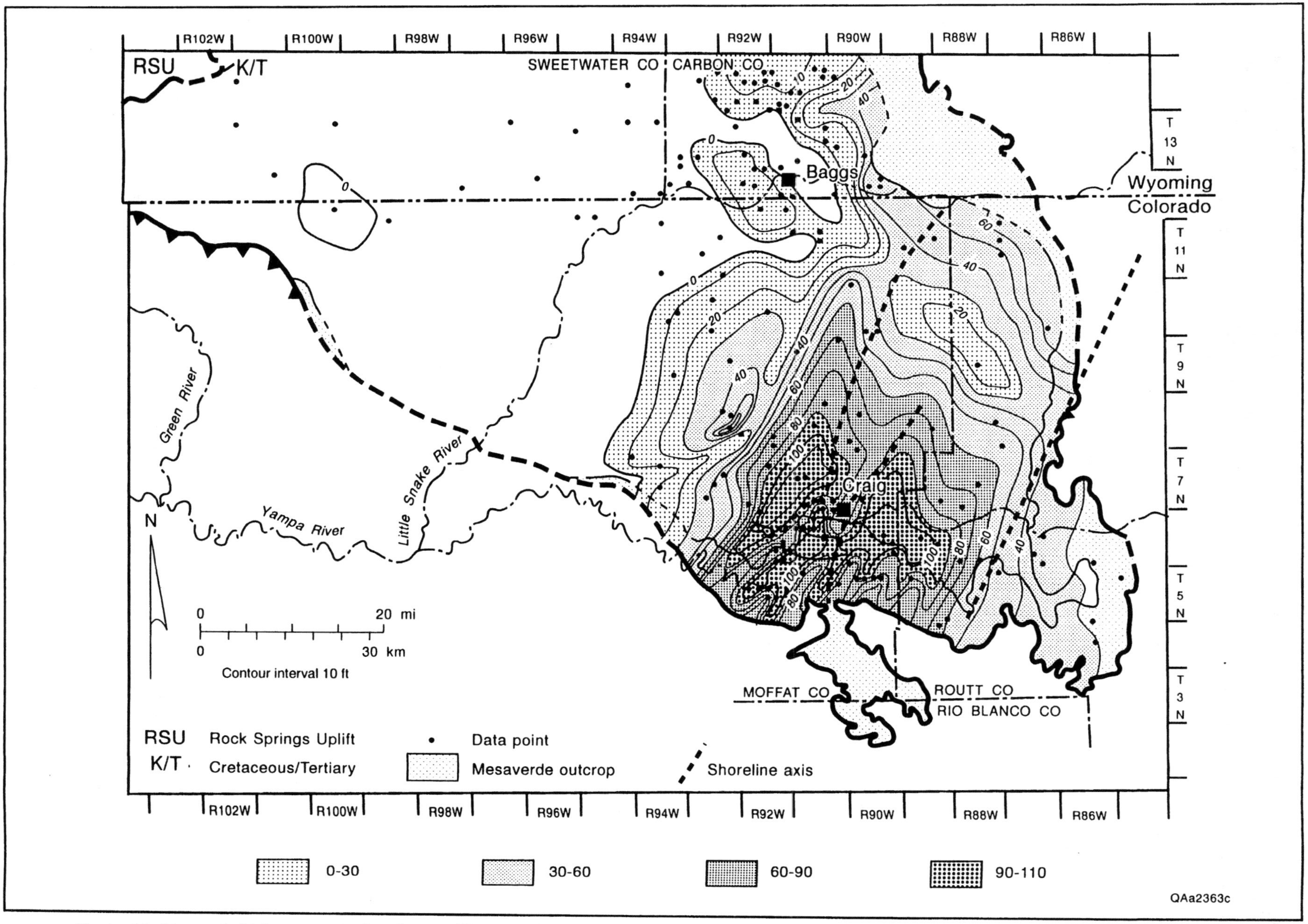

Figure 16. Net-coal-thickness map of Unit 1, Williams Fork Formation. Thickest net coal occurs in the Craig area, where peat accumulated on the coastal plain behind the linear shoreline systems. Coals thin to the west at the coastal-alluvial plain transition. Coal is also thin along a dip-oriented (northwest-southeast) distributary system extending southeastward from Baggs.

the coal packages are continuous from the subsurface to the eastern, northeastern, and southern outcrop belts. They are thus exposed to meteoric recharge and are potential conduits for basinward flow of ground water.

Geologic Controls on Coal Seam Occurrence

Peat accumulation and preservation as coal is dependent on three critical factors: (1) substantial growth of vegetation, (2) maintenance of the water table at or above the sediment surface, and (3) nondeposition of clastic sediment during peat accumulation. Substantial vegetation growth is mostly determined by climate, and the second two critical factors are controlled by the depositional systems, basin subsidence, and hydrology. The depositional systems provide the framework within which the peat swamps are established, and, combined with subsidence and hydrologic regime, are important in maintaining optimum water table levels for peat preservation.

Distribution of the Unit 1 coals is intimately related to the depositional systems and basin subsidence trends. Three salients (net coal greater than 100 ft [30 m]) are apparent on the net coal thickness map (fig. 16). Each lies immediately landward of successive strandplain axes of the linear shoreline system (compare figs. 15 and 16). The coastal plain is an area of sediment bypass and provides an opportunity for uninterrupted peat accumulation. The ideal location for preservation of the peat is immediately behind the shoreline system where water tables are maintained at optimum levels. Basin subsidence is also an important underlying control on coal occurrence. It determines the location of clastic sedimentation and accommodation space for peat accumulation. The Unit 1 coals are oriented northeast-southwest, which parallels the basin subsidence trend. The coals thin to the southeast and are ultimately limited by the final position of the shoreline, beyond which marine conditions existed.

Net coal thickness gradually thins westward at the transition between the coastal and alluvial plain systems. The alluvial plain probably resembled a piedmont surface that graded slowly down to the low-lying coastal plain. This surface gradient would have strongly influenced ground-water levels such that the water table was highest immediately behind the shoreline and progressively lower in the landward direction. Lowering of the water table is postulated to account for the gradual westward thinning of the coastal plain coals. Thick coals were not preserved toward the landward side of the coastal plain, despite there being a uniformly broad area bypassed by coarse clastic sediments (as defined by the 125 ft [38 m] contour; fig. 15).

Unit 1 coals are also thin along a narrow, dip-oriented zone that extends southeastward through Baggs (fig. 16). The coals are partially replaced by stacked sandstone units that are interpreted as distributary channels. These distributaries cut across the coastal plain and were the dispersal pathways for coarse clastic sediment delivered to the prograding shoreline system.

Unit 2

The second genetic depositional sequence of the Williams Fork Formation, Unit 2, is a clastic wedge similar to that of Unit 1, except that it did not prograde as far basinward. Unit 2 is bounded by regionally extensive, low-resistivity shale markers. The lower boundary is a flooding surface that terminates the coal-forming conditions of Unit 1 (fig. 11). The upper bounding surface is another maximum flooding surface that underlies the progradational Twentymile Sandstone. Unit 2 is characterized by upward-coarsening, progradational log patterns of the Sub-Twentymile Sandstone in the eastern and southeastern parts of the basin (Siepman, 1986). Log facies change to the west into aggradational, blocky channel-fills and interbedded mudstones of the upper Ericson Sandstone (Canyon Creek Member; fig. 12). Unit 2 is therefore stratigraphically equivalent to the Sub-Twentymile Sandstone sequence in the eastern part of the basin, and the Canyon Creek Member of the Ericson Sandstone on the southern flank of the Rocks Springs Uplift. Unit 2 ranges from 200 to 350 ft [61 to 107 m] thick, and basin subsidence is greatest in the southeast.

Depositional Systems

Depositional setting of Unit 2 is comparable to that of Unit 1, and three major depositional systems are recognized from the geometry of the framework sandstones and log facies mapping. The eastern part of the basin was characterized by a linear shoreline system that was backed landward by the coastal plain and farther landward by the alluvial plain.

The shoreline system is defined by two subparallel strike-oriented (northeast-southwest) sandstone-rich trends (fig. 17). The progradational character of this system is indicated by the prominent upward-coarsening log profiles of the sandstones (figs. 12 and 13). The Unit 2 shoreline system is similar to that of Unit 1 except that progradation did not extend as far basinward. The seaward-most sandstone-rich shoreline trend is crosscut by a sand-poor trend interpreted as a tidal-inlet complex. Farther northwest, the inlet complex passes into a dip-oriented distributary channel complex that cuts across the coastal plain. The coastal plain, landward of the shoreline system, is defined by net sandstone from approximately 75 to 150 ft (22.8 to 45.7 m). As in

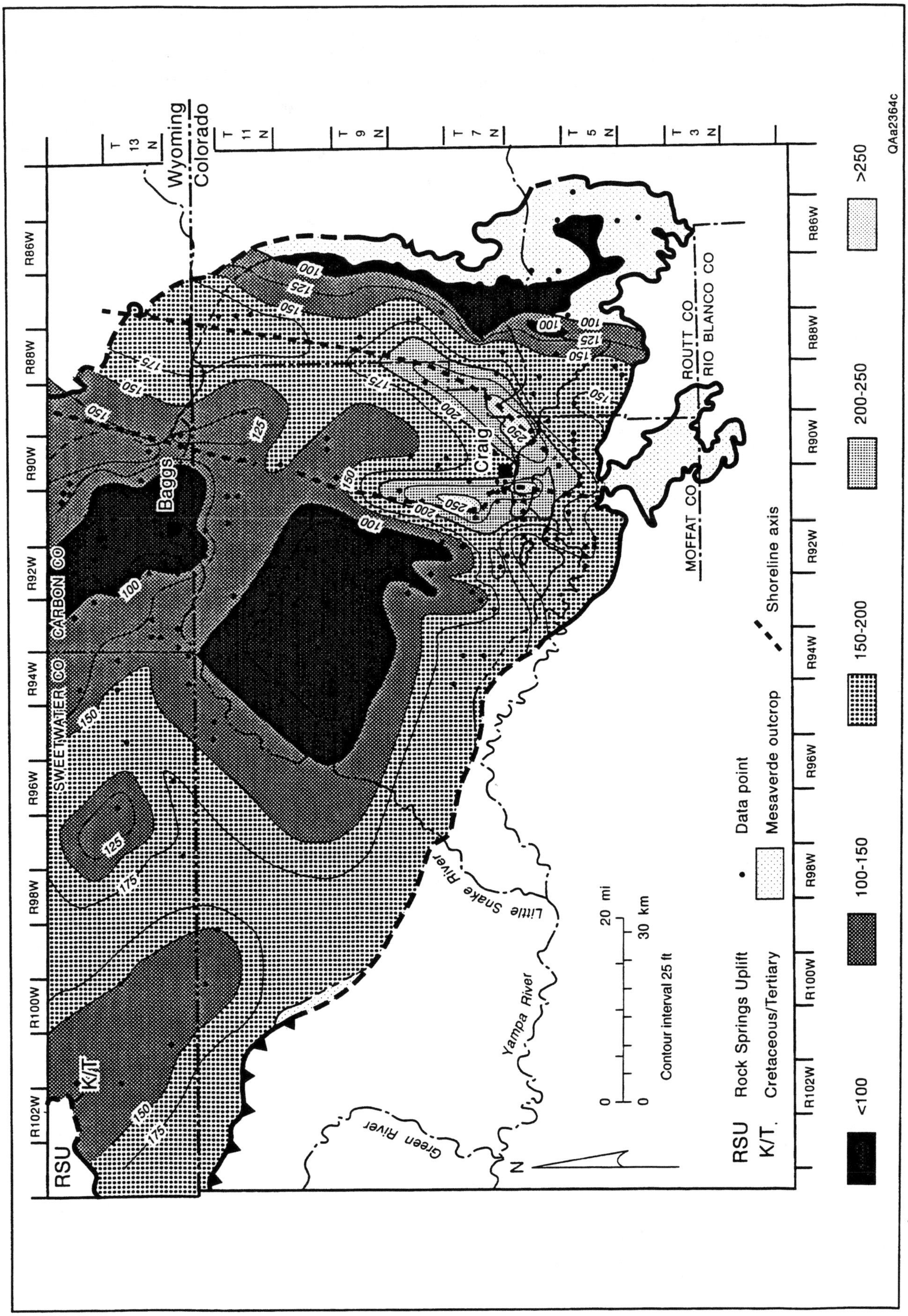

Figure 17. Net-sandstone map of Unit 2, Williams Fork Formation. Two subparallel, strike-oriented sandstone-rich trends define the shoreline system that is backed landward by a coastal plain system (net sandstone less than 150 ft [<46 m]), which grades westward into the alluvial plain. The sand-poor trend crosscutting the seawardmost shoreline is interpreted as a tidal-inlet complex. The sandstone-rich trend crosscutting the coastal plain represents a distributary channel complex.

Unit 1, the Unit 2 coastal plain was largely an area of coarse clastic sediment bypass, and log patterns are aggradational, reflecting thick coals and interbedded mudrocks. The coastal plain grades landward into the alluvial plain to the west. Sandstone trends are ill-defined on the alluvial plain, but the aggradational log patterns result from stacked channel-fills and interbedded muds. The alluvial plain was probably an elevated piedmont broadly traversed by a sandy bed-load fluvial complex.

Coal Stratigraphy

Unit 2 contains two coal seams that can be individually correlated over broad areas by their distinctive density and gamma-ray profile (fig. 14). The seams do not extend to the east as far as those of Unit 1 because they were limited by the extent of the Unit 2 progradational platform. Unit 2 was a minor progradational episode. The lower of the two coals (seam no. 3; fig. 14) is correlated throughout T6-8N; R92W and varies in thickness from 15 to 20 ft (4.5 to 6 m). The upper coal (seam no. 4; fig. 14) is correlated from the southern outcrop belt (T5N; R91-92W) to the northeastern outcrop (T13N; R90W) where it is equivalent to the Pioneer Coal in the Dixon field (fig. 13). This coal is as much as 25 ft (7.6 m) thick but splits locally into three seams, 5 to 15 ft (1.5 to 4.5 m) thick (fig. 18). The upper coal appears to be continuous as far east as T6N; R89W, where a 6-ft (1.8-m) seam is present. However, the seam character is not definitive where the coal is thin.

Coal Distribution

Net coal thickness within Unit 2 is at a maximum to the west and northwest of Craig, where it averages 40 ft (12.2 m) thick (fig. 19). There is a pronounced north-northeast/south-southwest alignment to coal-thickness trends. Net thickness decreases westward to less than 10 ft (3 m) along a line approximately defined by R94-95W. Thinning also occurs in the easternmost part of the basin where coal is absent beyond R87W, and along a narrow, northwest-southeast-trending zone from T13N; R93W to T9N; R89W (fig. 19). Although the coals are thinned along this trend, the net coal thickness map indicates that the coal-bearing packages are continuous from the subsurface to the northeastern and southern outcrop belts. Similar to Unit 1 coals, these coals are exposed to meteoric recharge and are potential conduits for basinward flow of ground water.

Geologic Controls on Coal Seam Occurrence

Geologic controls on Unit 2 coal distribution are comparable to those of Unit 1. The coals are thickest and most continuous on the coastal plain immediately landward of shoreline system (compare figs. 17 and 19). Isolation from sediments and maintenance of high water table levels provided by the coastal plain make it the optimum site for peat accumulation and preservation. Unit 2 coals trend north-northeast, which parallels the shoreline trend. Net coal thickness gradually thins to the southeast (fig. 19), reflecting increasingly marine-dominated deposition, and is limited ultimately by the seaward extent of the shoreline system. A cross section illustrating the relationship between the thick Unit 2 coals and the prograding shoreline system is shown in figure 20. Peat accumulation is greatest on the aggradational coastal plain. The peats can override the shoreline sandstones to achieve greater lateral extent, but are thinner.

As was the case in Unit 1, the Unit 2 coals also thin gradually westward and are lost just beyond the transition between the coastal and alluvial plain systems. Gradual westward thinning of the coastal plain coals is again thought to be the result of a lowering water table with increased gradient on the piedmont surface of the alluvial plain system.

Thinning of the Unit 2 coals along the narrow, west-northwest/east-southeast-trending zone from T11N; R93W to T10N; R88W overlaps the distributary channel and tidal-inlet complex illustrated on the net-sandstone map (fig. 17). Peat accumulation was probably inhibited along this zone by clastic deposition in the distributary complex and by marine influence associated with the tidal complex.

Unit 3

The third genetic depositional sequence of the Williams Fork Formation, Unit 3, is a clastic wedge that extended shoreline and coastal plain deposits farther basinward than Unit 2, but not as far as Unit 1. Unit 3 is also bounded by regionally extensive, low-resistivity shale markers. The lower boundary is the maximum flooding surface that precedes the Twentymile Sandstone progradation (figs. 11 through 13). The upper boundary represents a minor transgressive event, and the facies offset above this marker is subtle (fig. 12). Unit 3 is dominated by the upward-coarsening and blocky log profiles of the Twentymile Sandstone over the eastern half of the basin. To the west, the log facies change to mud-rich aggradational patterns (fig. 12). Stratigraphically, Unit 3 includes the Twentymile Sandstone and overlying coals in the east, and the lower part of the Almond Formation (as defined at the Rock Springs Uplift by Roehler, 1987) in the west. Unit 3 is also equivalent in part to the Pine Ridge Sandstone to the north. Thickness of the unit varies from 200 ft (61 m) in the northeast to 450 ft (137 m) in the southeast.

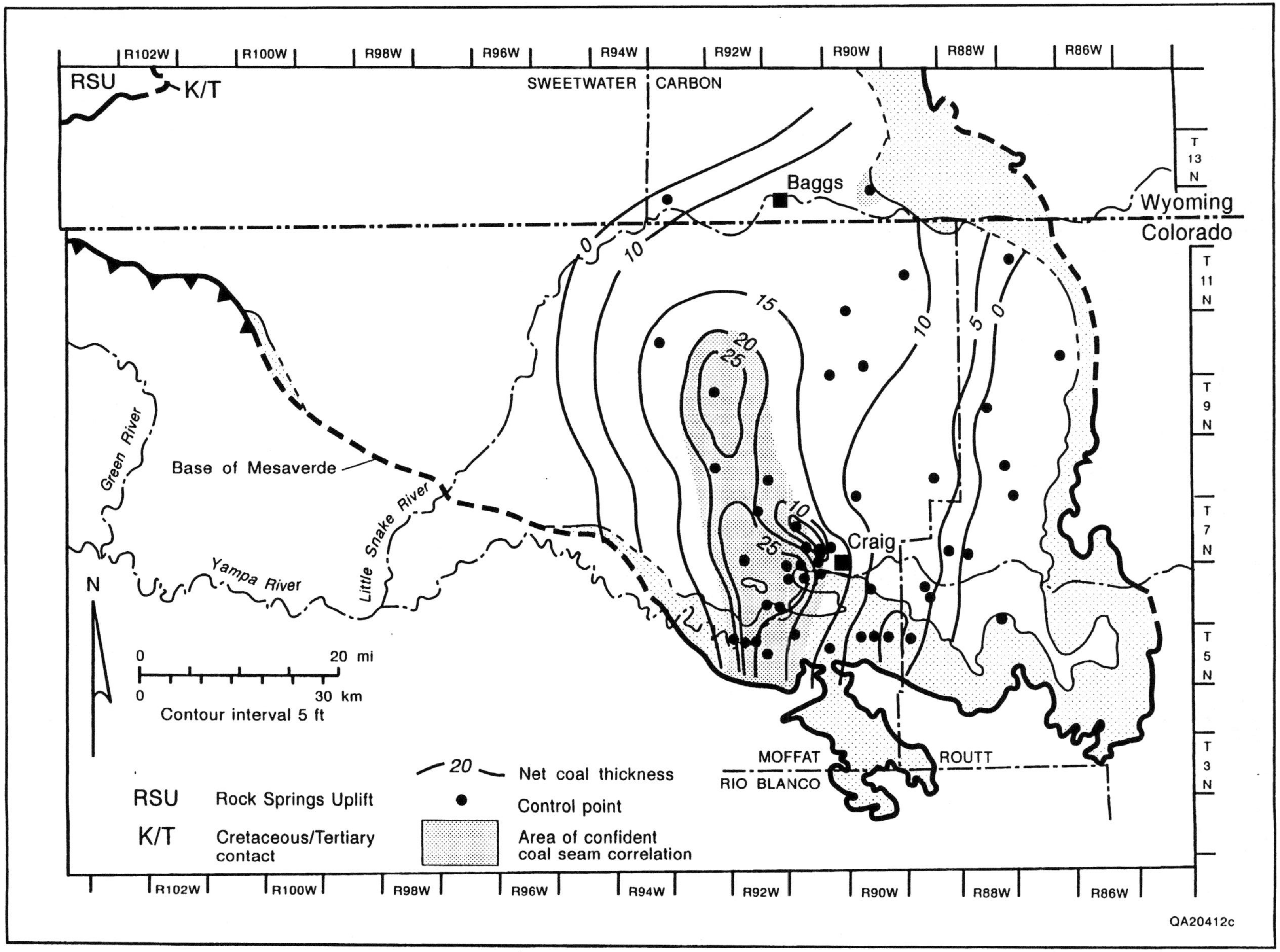

Figure 18. Isopach of the Unit 2 coal seam. The seam trends north-south, and is continuous to the northern and southern outcrop belts, where it is exposed for recharge. Peat accumulated on the coastal plain behind a linear shoreline system.

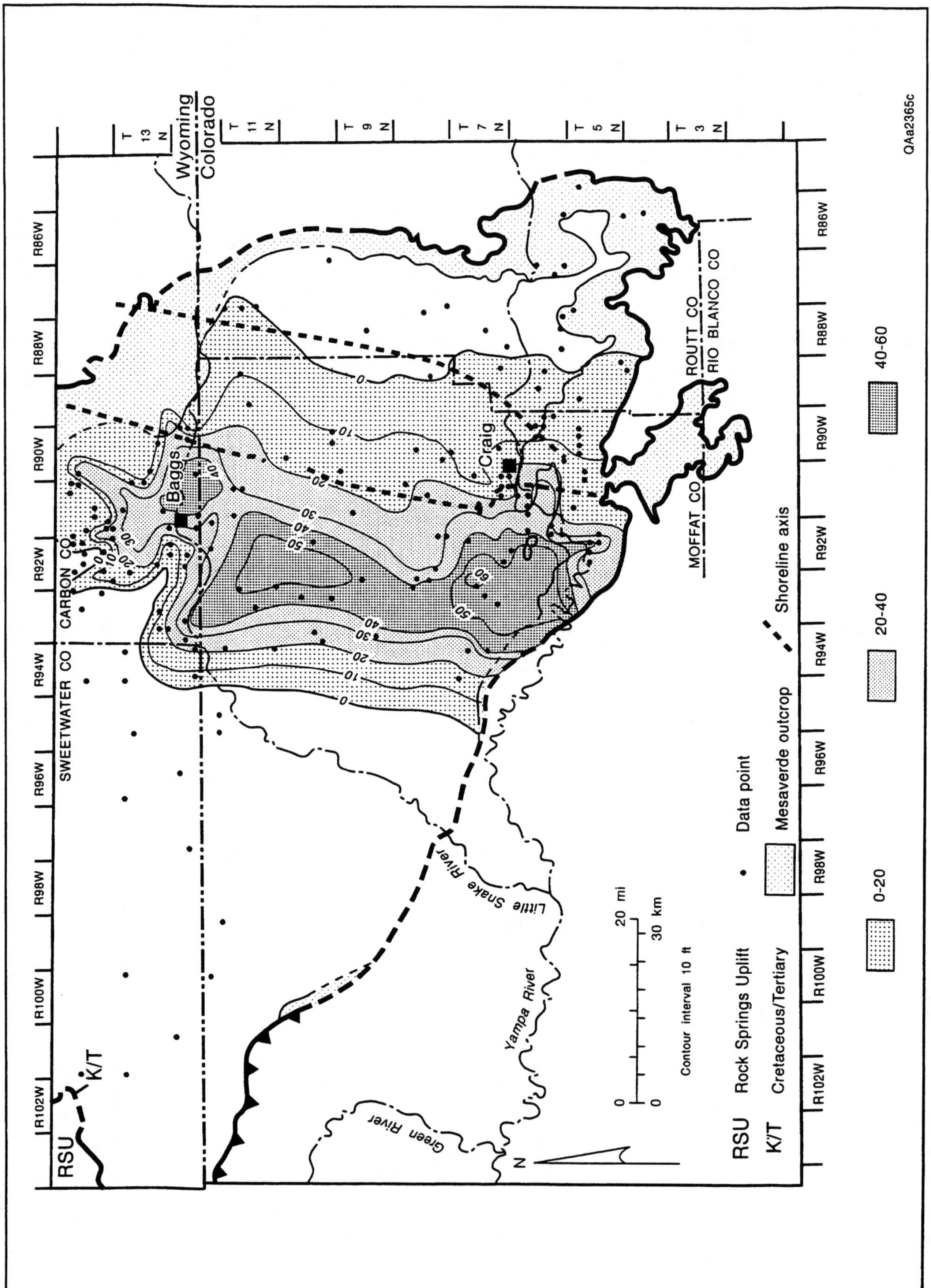

Figure 19. Net-coal-thickness map of Unit 2, Williams Fork Formation. Thickest net coal occurs west and northwest of Craig, where peat accumulated on the coastal plain behind the linear shoreline systems. Coals thin to the west at the coastal-alluvial plain transition. Coal is also thin along a dip-oriented (northwest-southeast) trend that overlaps the distributary channel and tidal-inlet complex.

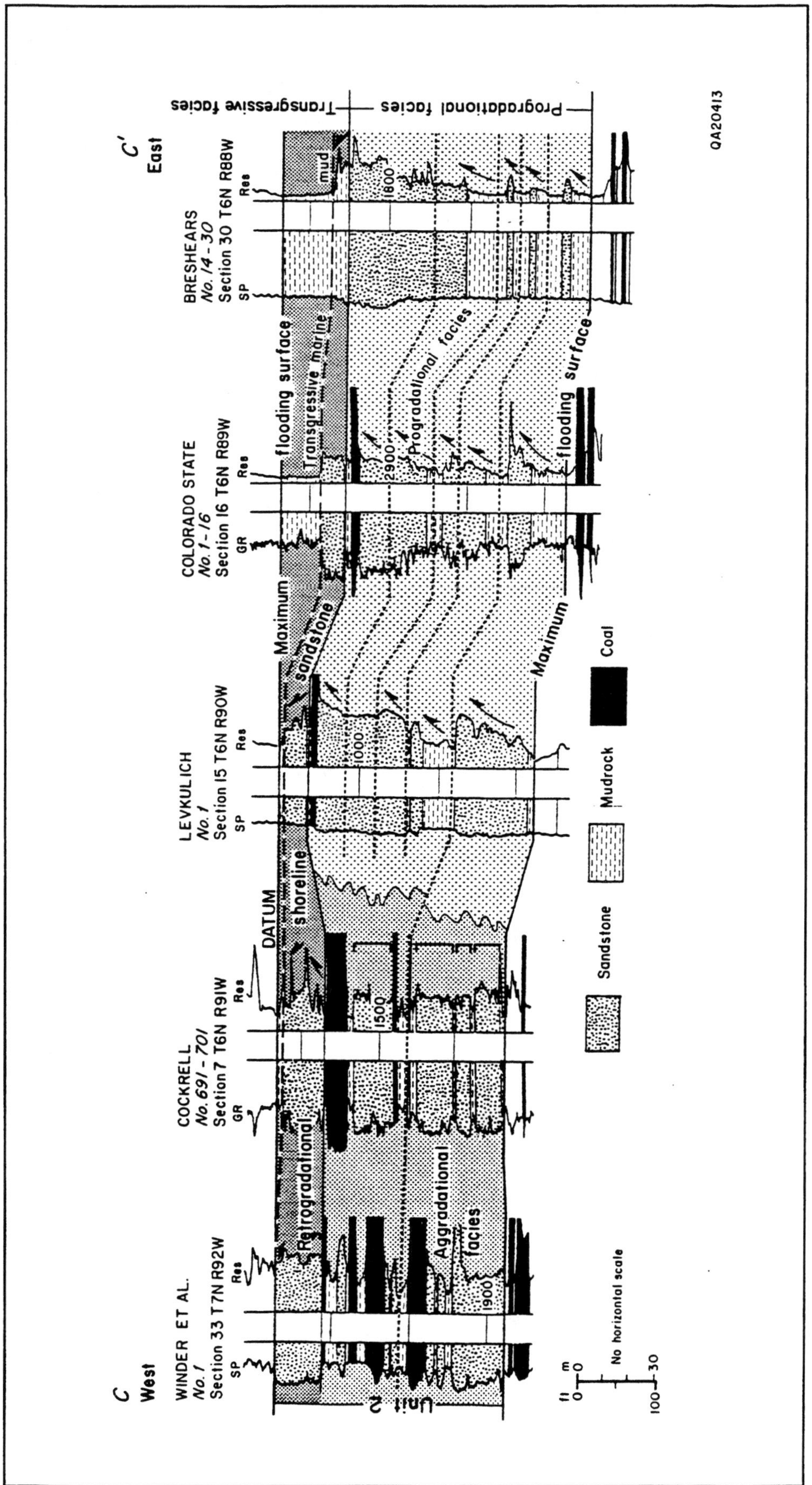

Figure 20. Cross section of the Unit 2 coal. Peat accumulated on an aggrading coastal plain behind the progradational shoreline facies. Coals override the shoreline sandstones to achieve great lateral extent.

Basin subsidence trends continue to show northeast-southwest alignment.

Depositional Systems

Sandstone geometry and log facies distribution indicates that Unit 3 was deposited in a setting similar to underlying Units 1 and 2. Two parallel (north-northeast oriented) sandstone-rich trends occupy the southeastern part of the basin and define successive linear shorelines (fig. 21). The blocky/upward-coarsening log motifs of the shorelines rise stratigraphically to the southeast, suggesting progradation of the shoreline system. Landward of the shoreline system is a coastal plain system defined by several large sandstone-poor areas that are characterized by aggradational log facies consisting of mudstone-rich, coal-bearing deposits. The sandstone-poor areas are separated by broad, dip-oriented sandstone-rich belts representing moderately sinuous mixed-load channels that cut across the coastal plain (fig. 21). The channel fills display prominent blocky and upward-fining log facies.

The coastal plain passes gradually westward (landward) into the alluvial plain and, although data are scarce and the percent-sandstone map is not definitive, log facies distribution suggests that deposition occurred in a mixed-load fluvial system. Log motifs are dominated by mudstone-rich aggradational packages with minor interbedded blocky and upward-fining channel units (fig. 12).

Coal Stratigraphy

Coals associated with Unit 3 can be mapped only in broad packages and are thin, except in the vicinity of T6-8N; R91-93W, where as many as three seams ranging from 2 to 8 ft (0.6 to 2.4 m) thick are present, and in T13-14N; R91-93W, where as many as six seams ranging from 2 to 20 ft (0.6 to 6 m) thick are present. Downdip, the coal package consists of three to six seams, and although only 2 to 3 ft (0.6 to 0.9 m) thick, the coals extend to the limit of well data (T5N; 89W) and probably continue to the southeastern outcrop belt.

Coal Distribution

Unit 3 coal packages are extensive over the eastern half of the basin and attain a maximum thickness north of Baggs, where the coals are as much as 48 ft (14.6 m) thick. Elsewhere, the net coal thickness contours define several discrete coal lenses that are typically greater than 30 ft (9.1 m) thick. The coal lenses trend strongly in the strike (northeasterly) direction, but there are dip-oriented lenses to the northwest of Baggs and between T12N; R98W and T13N; R101W (fig. 22). Net coal thickness decreases to the east, and coals are absent beyond R87W. The coals also thin to the west and are generally thinner than 10 ft (3 m) west of the Little Snake River. The coals are exposed in the southern and northeastern outcrop belts.

Geologic Controls on Coal Seam Occurrence

Geologic controls on Unit 3 coal distribution are very similar to those of Units 1 and 2. The coals are strongly strike-oriented and are thickest and most continuous on the coastal plain immediately landward of the shoreline system (compare figs. 21 and 22). Similar to Units 1 and 2, the coastal plain maintained high water table levels and was isolated from much of the sediment load, thus providing an optimum site for peat accumulation and preservation. However, Unit 3 coals are not as extensive as Unit 1 and 2 coals because the area of sediment bypass was not as great and subsidence rates were perhaps not as favorable. The Unit 3 coastal plain was more obviously dissected by mixed-load fluvial channels than occurred in Units 1 and 2. Unit 3 coals thin gradually westward and are lost just beyond the transition between the coastal and alluvial plain systems. Gradual westward thinning of the coastal plain coals, as in Units 1 and 2, is thought to be the result of a lowering water table with the increased gradient of the alluvial plain. Thinning of Unit 3 coals to the east and southeast is a result of increased marine conditions and the seaward extent of the shoreline system defines the easterly limit of coal distribution (fig. 21).

Unit 3 coals are dip oriented near Baggs, where the mixed-load fluvial system cuts across the coastal plain. Coal distribution along this trend is controlled by the fluvial system, as evidenced by the thin coals along the channel axes that thicken toward the interchannel areas. Unit 3 coals in the western half of the basin are also dip oriented, probably because of fluvial control, but well coverage is too sparse to demonstrate the relationship.

Unit 4

The uppermost genetic depositional sequence of the Williams Fork Formation is Unit 4. It is characterized throughout by aggradational, mudstone-rich coal-bearing deposits overlying a very thin progradational base (figs. 11 through 13). Thus, the facies offset from underlying mudstone-rich coal-bearing rocks of Unit 3 is subtle. The flooding event that defines the base of Unit 4 was minor when compared to the other flooding

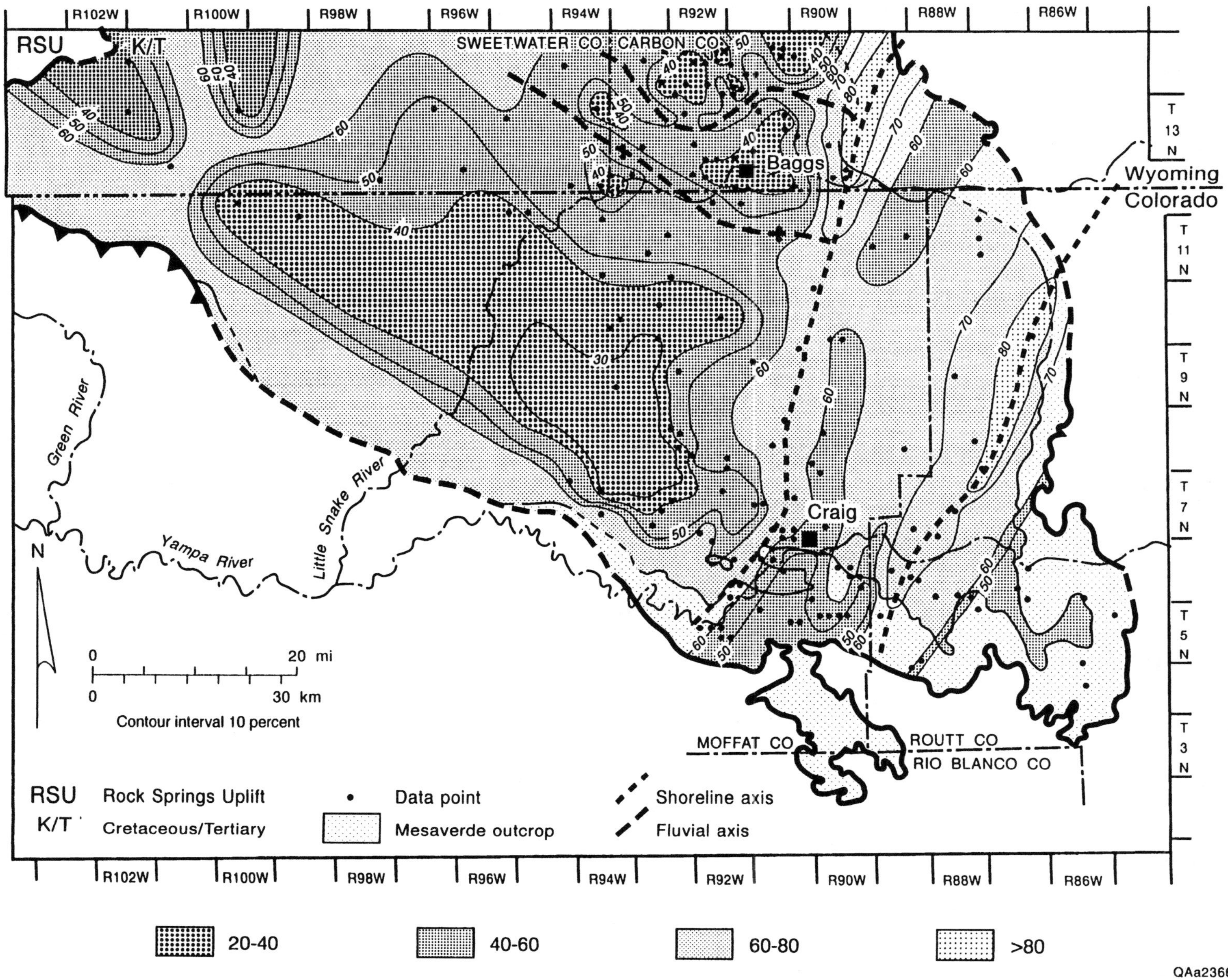

Figure 21. Percent-sandstone map of Unit 3, Williams Fork Formation. Two parallel, strike-oriented sandstone-rich trends define the shoreline system that is backed landward by a coastal plain system, which in turn grades westward into the alluvial plain. The coastal plain is cut by a number of broad, dip-oriented sandstone-rich belts representing moderately sinuous mixed-load channels.

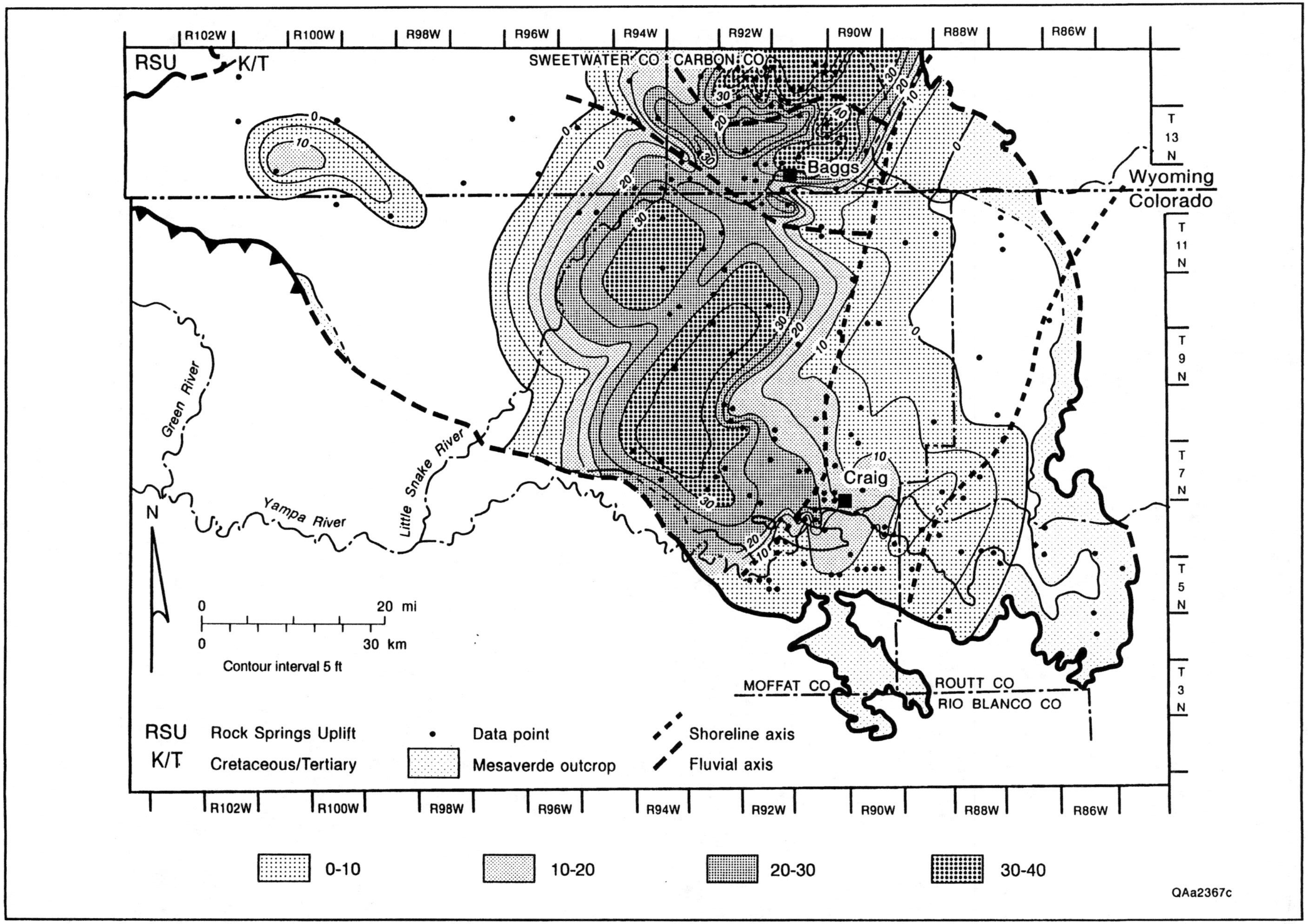

Figure 22. Net-coal-thickness map of Unit 3, Williams Fork Formation. Thickest net coal occurs west and northwest of Craig, where peat accumulated on the coastal plain behind the linear shoreline systems. Coals thin to the west at the coastal-alluvial plain transition and are also thin along broad, dip-oriented (northwest-southeast) trends that were cut by fluvial channels.

surfaces that punctuate the Williams Fork sequence. The upper bounding surface separates the Williams Fork Formation from the overlying Lewis Shale in the northeastern half of the basin, and upward-coarsening barrier/strandplain facies of the genetically defined Almond Formation in the west and southwest (fig. 12).

Depositional Systems

A low to moderate sinuosity, mixed-load fluvial system dominated Unit 4 deposition, which is in contrast to the underlying genetic units. The percent-sandstone map defines several well-integrated sandstone-rich channel belts that are separated by extensive sandstone-poor floodplain areas (fig. 23). The channel belts generally trend southeasterly but can swing to the northeast in broad meanderloops and merge along the eastern edge of the basin with a north-south-oriented shoreline system. Log character within the sandstone-rich belts changes from blocky and blocky/upward-fining in the proximal, northwesterly facies to upward-fining in the southeast as the channels approached the shoreline system. The floodplain areas are characterized by aggradational, mudstone-rich coal-bearing deposits and thin, upward-coarsening sequences indicating possible lacustrine influence. The shoreline system is characterized by upward-coarsening log profiles of thick, stacked shoreface sandstones.

Coal Stratigraphy

Coals of Unit 4 can be correlated throughout the eastern part of the Sand Wash Basin as two broad groups (fig. 12), although individual seams can be correlated locally. The lower group consists of two to five seams, 1 to 15 ft (0.3 to 4.5 m) thick, whereas the upper group consists of two to four coals 1 to 8 ft (0.3 to 2.4 m) thick. Coals tend to be thicker west of R90W. The coals extend as far as the eastern limit of the area defined by well information and appear to project to the eastern outcrop belt, although they thin to the east.

Coal Distribution

In the eastern half of the Sand Wash Basin, Unit 4 coals are distributed in isolated pods that range from small (10 to 30 mi^2 [2.5×10^7 to 7.7×10^7 m^2]) to large (100 to 150 mi^2 [2.58×10^8 to 3.87×10^8 m^2]). The pods tend to be elongate and dip-oriented, but considerable variability is apparent on the net-coal-thickness map (fig. 24). The thickest net coal occurs in a trend of pods that extends northwestward from the southern outcrop belt near Craig. Net coal thickness along this trend averages 40 ft (12.2 m), but it can be as much as 53 ft (16.1 m). Near Baggs, net coal thickness is typically 20 to 30 ft (6 to 9.1 m). The Unit 4 coals are exposed along the southern and northeastern outcrop belts, but they do represent a continuous interconnected aquifer system in the subsurface.

In the western half of the basin, the coals occur in a strongly dip-oriented trend (southeasterly) that extends from the southern edge of the Rocks Springs uplift to the Little Snake River. The trend is only broadly defined because well control is sparse. Net coal thickness of this trend averages 25 ft (7.6 m) and represents the first major Williams Fork coal occurrence west of the Little Snake River. Although the coals are exposed along the southern edge of the Rocks Springs Uplift, they are substantially thinned at the outcrop and may not be efficient conduits for basinward flow of ground water.

Geologic Controls on Coal Seam Occurrence

Unit 4 coal distribution and thickness are controlled by the mixed-load fluvial system. Comparison of the percent-sandstone and net-coal-thickness maps (figs. 23 and 24) indicates that the isolated pods of thick coal (net coal thickness from 30 to 40 ft [9.1 to 12.2 m]) are located in the sandstone-poor interchannel areas. Conversely, the coals are thin (net coal thickness typically less than 15 ft [4.5 m]) along the sandstone-rich channel belts. A more subtle trend also evident from the maps is that there is an optimum distance from the channel belts for peat accumulation and preservation, particularly in the downstream reaches of the fluvial system. The coals are thicker adjacent to the channel belts and thin with increased distance from the channels. This relationship does not hold in the thick coal trend to the west and northwest of Craig. However, two major channel belts converge in this area, suggesting that no point in the interchannel peatswamp was very far from a channel complex. The ideal location for accumulation and preservation of the peat may have been in the interchannel areas between major fluvial axes, but not too far into the floodplain where greater subsidence resulted in lacustrine inundation.

Almond Genetic Depositional Sequence

The Almond genetic depositional sequence is characterized by a series of upward-coarsening sandstone-rich cycles that are separated by thin, mudstone-rich, coal-bearing units. The lower bounding surface is an extensive, low-resistivity shale marker, and the upper bounding surface is defined by the overlying Lewis Shale (fig. 12). The prominent change

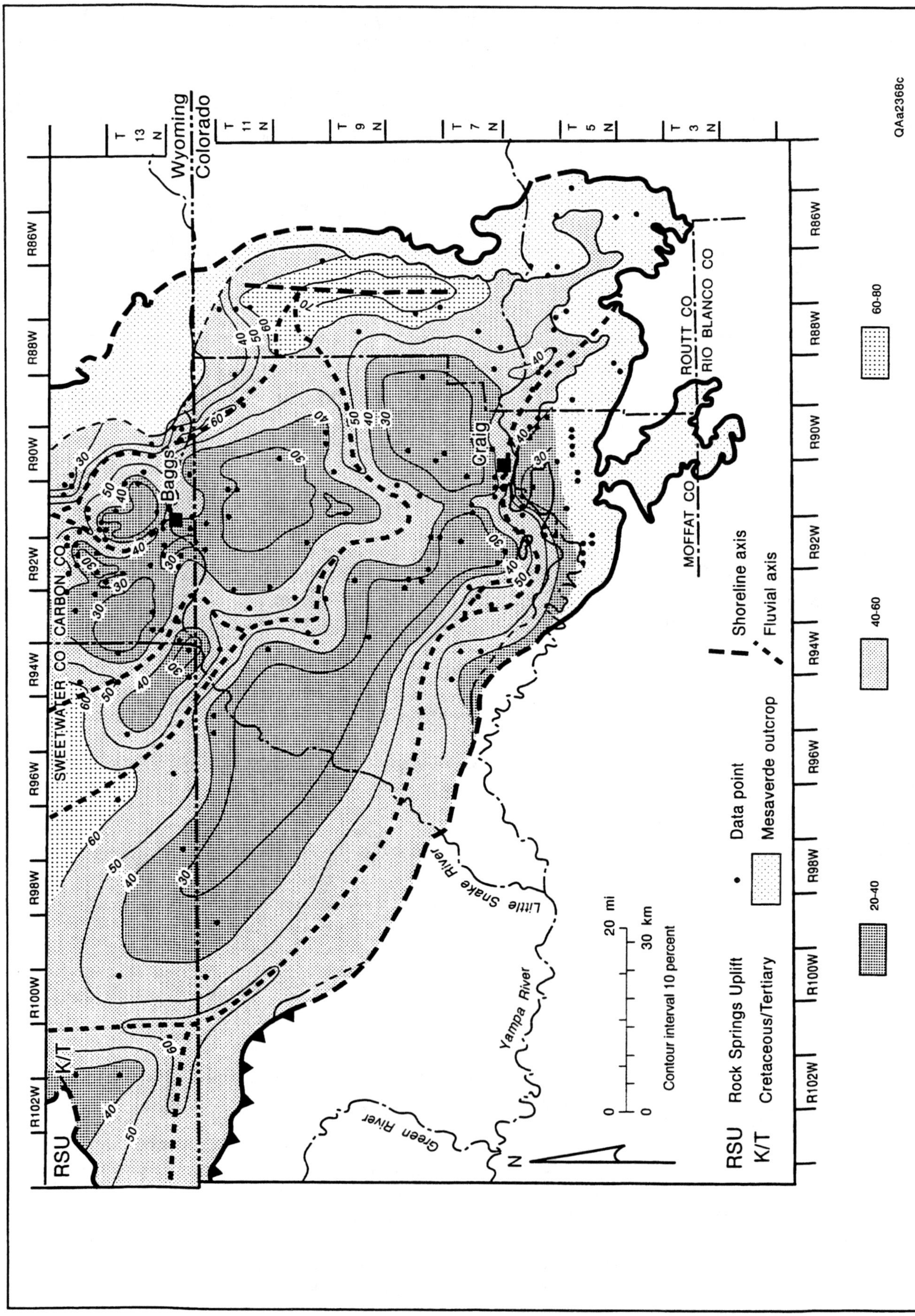

Figure 23. Percent-sandstone map of Unit 4, Williams Fork Formation. The sandstone map defines several well-integrated sandstone-rich channel belts displaying moderate sinuosity. The channel belts are separated by sandstone-poor interchannel areas and merge in the southeast with a linear shoreline.

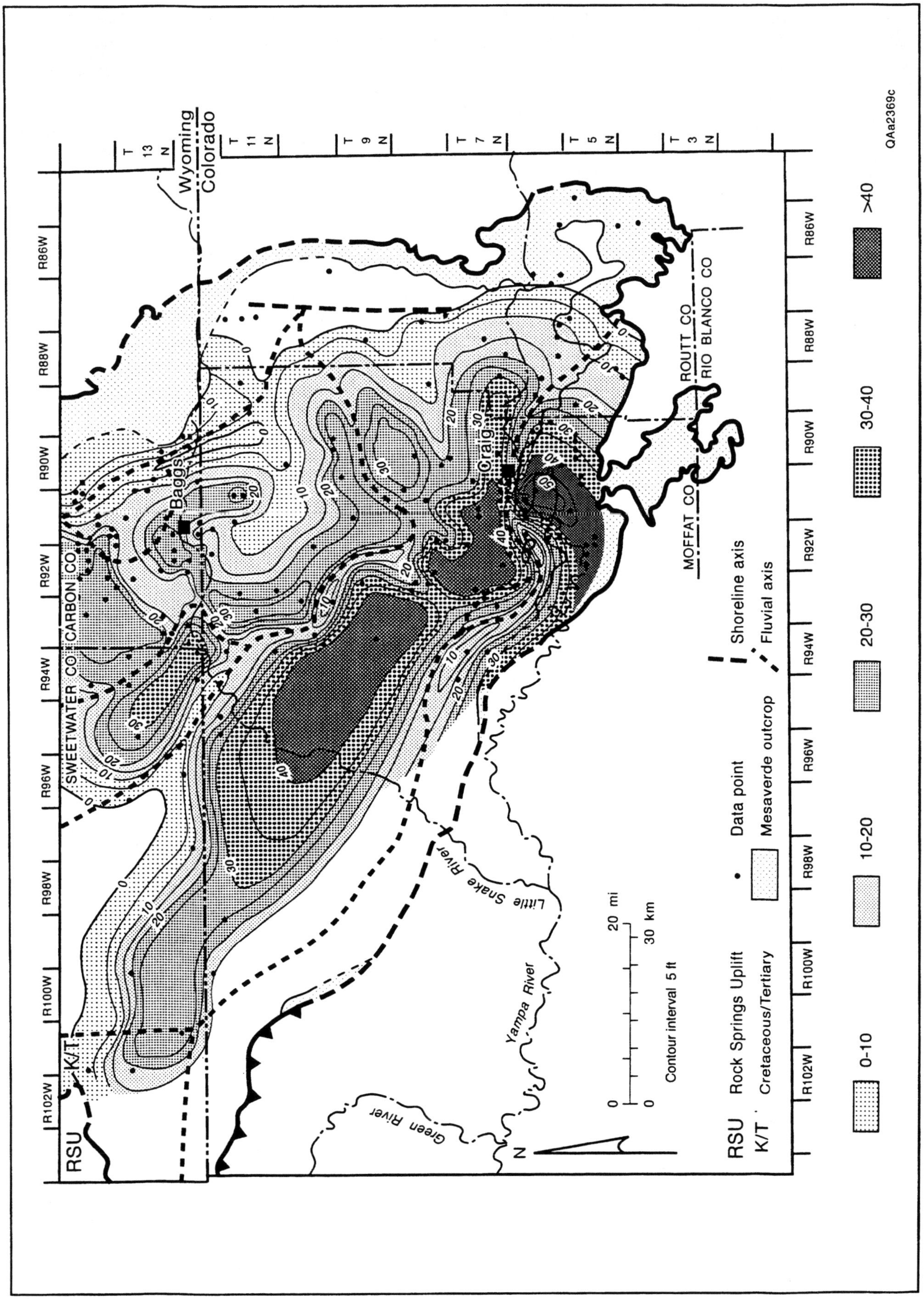

Figure 24. Net-coal-thickness map of Unit 4, Williams Fork Formation. The coals occupy the areas adjacent to major channel belts. Thickest net coal occurs along a dip-oriented trend of isolated pods extending northwestward from the Craig area.

in gamma-ray, spontaneous potential, and resistivity log responses between the uppermost Williams Fork Formation and the Almond genetic unit facilitates recognition of the lower bounding surface. The Almond Formation was only deposited over the southwestern half of the basin. To the northeast, the Williams Fork Formation is directly overlain by the Lewis Shale. Thickness of the Almond unit varies across the Cedar Mountain Fault system, indicating that this fault zone was active during Almond deposition.

Depositional Systems

The Almond genetic sequence was deposited in a wave-dominated delta system. The percent-sandstone map (fig. 25) displays both strike-oriented and dip-oriented elements. Dip-oriented delta distributaries trend northeastward (indicating a southwesterly sediment source) and supplied sediment for wave reworking along strike into a strandplain system. The core of the strandplain averages 70 to 80 percent sandstone and extends northwestward from Craig. The strandplain grades landward (southwest) into a sandstone-poor (less than 50 percent) coastal plain. The strandplain ends abruptly in the seaward direction beyond which marine conditions prevailed.

Coal Stratigraphy and Distribution

Data for analysis of the coals are scarce and no detailed work has been attempted. The coals are typically thin and individually average from 2 to 5 ft (0.6 to 1.5 m) thick. Their continuity has not been demonstrated. The net-coal-thickness map (fig. 26) highlights three areas where net coal thickness is from 15 to 25 ft (4.5 to 7.6 m). The largest area is west of Craig, where coals trend in a northwesterly direction and net coal is at a maximum (as much as 25 ft [7.6 m]). The other areas of thick net coal are southeast of the Rocks Springs Uplift, where the coals trend northwestward, and west of the Sweetwater-Carbon County line, where coals are oriented to the northeast (fig. 26).

Geologic Controls on Coal Seam Occurrence

Comparison between the net-coal-thickness and percent-sandstone maps (figs. 18 and 19) suggests two relationships among coal distribution and depositional setting. The northwest-oriented coals correspond to low sandstone percentage and occupy a coastal plain position behind the core of the standplain system. Control on these coals is similar to that in Williams Fork Unit 1 through 3 coals. The coals to the west of the Sweetwater-Carbon County line are oriented northeasterly and lie in an area of low sandstone percent adjacent to a major delta distributary. Peat growth in this setting was initiated on the stable subdelta platform constructed by the distributary channel complex and maintained through the freshwater discharge delivered by the distributary complex.

Conclusions

1. The Williams Fork Formation, the most important coal-bearing unit in the Sand Wash Basin, can be divided into four genetic depositional sequences. These sequences were deposited during discrete episodes of basin history and are bounded by regionally extensive, low-resistivity shale markers that represent marine flooding surfaces in the basinward direction and hiatal, nondepositional surfaces in terrestrial facies (surfaces of sediment starvation). The Almond Formation is a minor coal-bearing unit and represents another depositional episode in the basin's history.

2. The first genetic depositional sequence, Unit 1, is a clastic wedge that extended coal-bearing coastal plain deposits beyond the present-day basin margin. Three depositional systems are recognized in the unit. A linear shoreline system dominated the easternmost part of the basin and was backed landward by a coastal plain system, which in turn graded westward into an alluvial plain system. Units 2 and 3 are clastic wedges displaying a similar arrangement of depositional systems, but these units did not prograde as far basinward. Unit 4 deposition was markedly different from the underlying units and was dominated by a low- to moderate-sinuosity mixed-load fluvial system. The Almond Formation was deposited as a wave-dominated delta system and is characterized by a large strandplain system stretching between delta distributaries.

3. Units 1 and 2 contain the thickest, most laterally extensive coals. Coal occurrence in all units is concentrated in the eastern half of the basin, and, with the exception of Unit 4, there is no significant coal to the west of the Little Snake River. Unit 1 and 2 coals are thickest in the vicinity of Craig, where net coal thickness averages 90 and 40 ft (27.4 and 12.2 m), respectively. Unit 3 coals are thickest northwest of Craig (average 30 ft [9.1 m]) and north of Baggs (average 40 ft [12.2 m]), and thickest Unit 4 coals (average 40 ft [12.2 m]) occur in a trend of isolated pods that extends northwesterly from the outcrop belt near Craig. Almond coals occur in three discrete areas where net coal thickness averages from 15 to 25 ft [4.5 to 7.6 m]. Continuity of the Williams Fork coals is variable. Some individual seams, particularly in Units 1 and 2, were

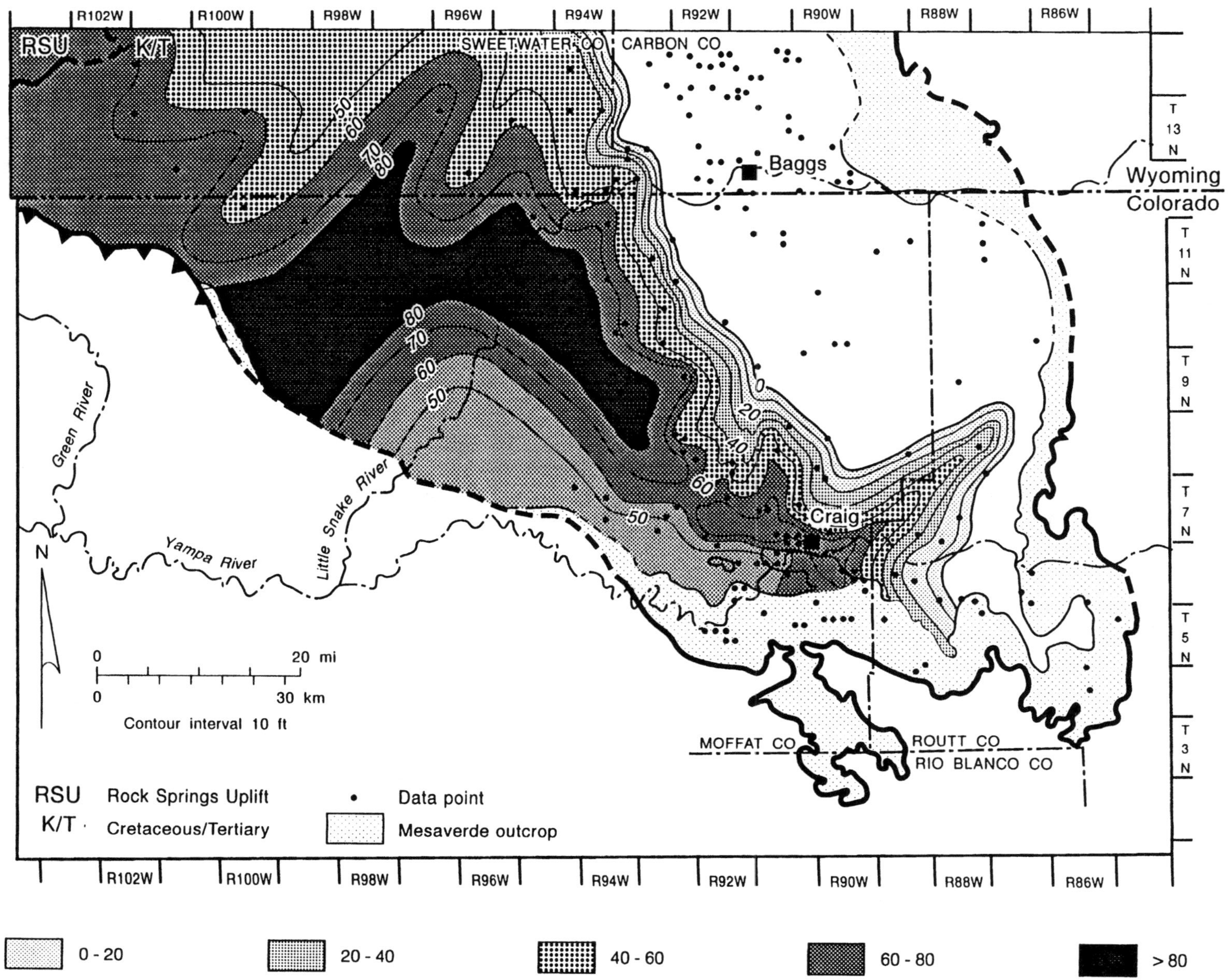

Figure 25. Percent-sandstone map of the Almond Formation. The map defines a strike-oriented strandplain system that extends between deltaic distributaries.

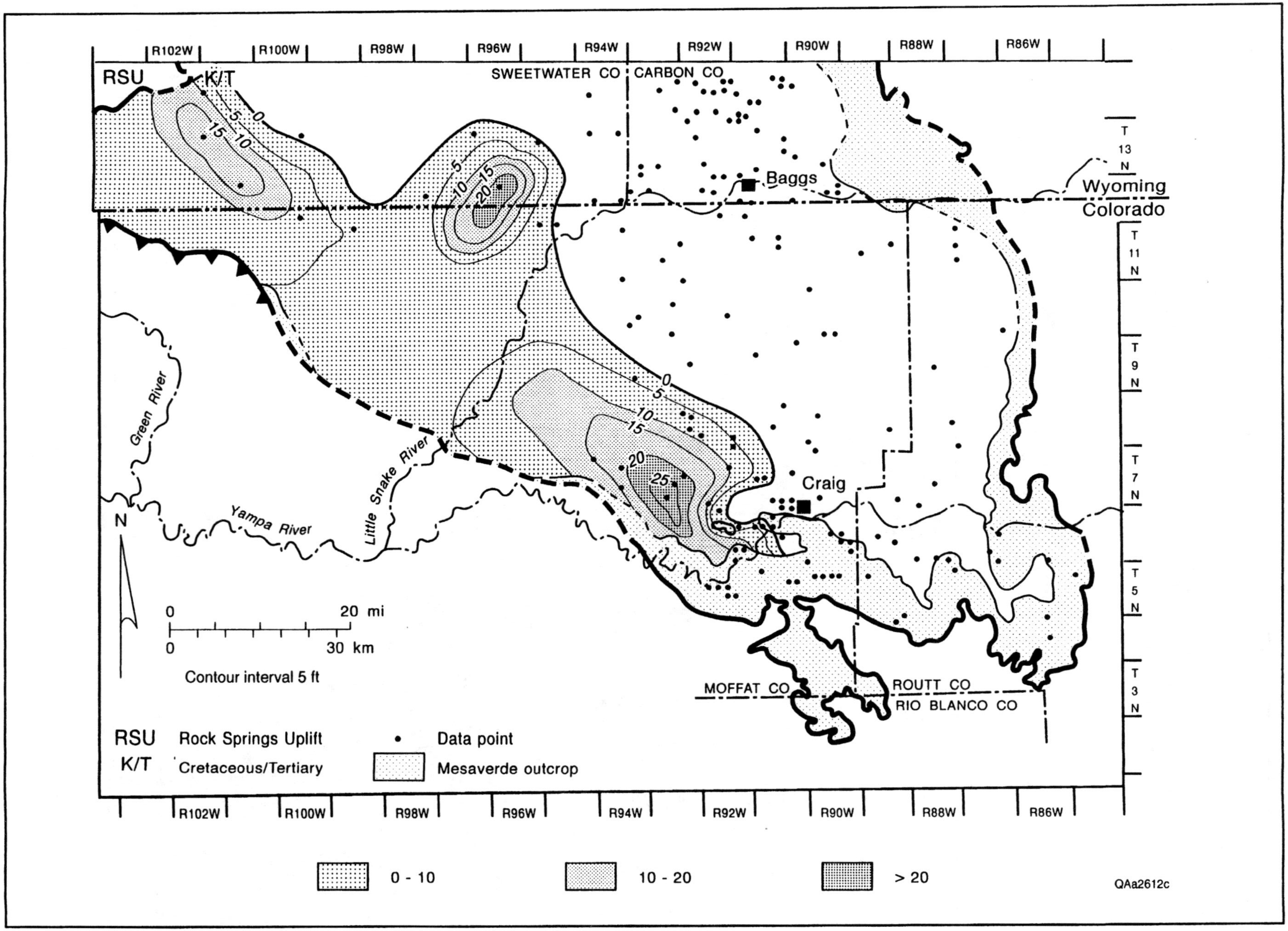

Figure 26. Net-coal-thickness map of the Almond Formation. Coals are distributed in three areas: (1) west of Craig, (2) southeast of the Rock Springs Uplift, and (3) west of the Sweetwater/Carbon county line.

correlatable throughout the eastern half of the basin by their characteristic profile on density and gamma-ray logs. Other seams could only be correlated when grouped as broad coal packages. The coals of Units 1 and 2 are continuous in the subsurface to the southern and northeastern outcrop belts and are exposed for meteoric recharge. Unit 3 and 4 coals are less continuous and are unlikely to represent interconnected aquifer systems in the subsurface. Continuity of Almond coals has not been demonstrated.

4. Coal occurrence in all units is intimately related to the depositional systems. The coastal plain immediately landward of the shoreline system was the optimum site for peat accumulation and preservation in Williams Fork Units 1 through 3 and the strike-oriented Almond coals. This was an area of sediment bypass and maintenance of optimum water table levels. Lowering of the water table is thought to account for the gradual westward thinning of the Williams Fork coals at the coastal plain/alluvial plain transition. Williams Fork Unit 1 through 3 coals override the shoreline sandstone to the east, but they also thin in this direction, and their ultimate lateral extent was limited by the final shoreline position, beyond which marine conditions prevailed. Unit 4 coals were preserved in an interchannel position between the channel axes of a mixed-load fluvial system. Dip-oriented Almond coals apparently accumulated adjacent to a major delta distributary that not only provided the platform for initiating peat growth but also supplied fresh water to maintain the peat swamp.

Coal Rank, Gas Content, and Composition and Origin of Coalbed Gases, Mesaverde Group, Sand Wash Basin

Andrew R. Scott

Abstract

Mesaverde coal rank ranges from high-volatile C bituminous along the basin margins to medium-volatile bituminous in the deeper parts of the basin. Coal rank in the eastern half of the basin, where the thickest coal beds occur, is generally high-volatile C to B bituminous. Mesaverde gas contents range from less than 1 to more than 540 scf/ton (<0.1 to >16.9 m^3/t) but are generally less than 200 scf/ton (<6.3 m^3/t). Gas contents change vertically within a well and laterally between wells. Factors controlling the distribution of high gas contents in the basin include coal rank, coal characteristics, localized pressure variations, basin hydrodynamics, and conventional trapping of migrating thermogenic and biogenic gases. Coalbed gases range from very wet to very dry (average C_1/C_{1-5} value of 0.96) but generally fall between C_1/C_{1-5} values of 0.94 to 0.99. Carbon dioxide content is variable, ranging from less than 1 to more than 25 percent. Coalbed gases are probably early thermogenic and secondary biogenic.

Thermal Maturity

The thermal maturity of coal-bearing basins is one of several factors that are important in determining the types and quantities of gases generated from coal beds. Coal is unusual because it acts as both the source of gas and the reservoir in which the generated gas is stored. Significant quantities of methane will be generated from coal once the threshold of thermogenic gas generation has been reached at approximately 0.80 to 1.00 percent vitrinite reflectance (Tang and others, 1991). Gas contents for higher-rank coal beds may exceed 400 to 500 scf/ton (>12.5 to 15.7 m^3/t) (Scott and Ambrose, 1992).

Secondary biogenic and early thermogenic coalbed gases are associated with low-rank coals that have not reached the threshold of thermogenic gas generation (Scott, 1993). Primary biogenic gases generated during peatification are probably not retained by the coal in significant quantities (Scott, 1993), whereas secondary biogenic gases are generated by bacteria introduced into the coals by meteoric waters flowing basinward from a recharge area. Gas contents associated with secondary biogenic methane generation are usually less than 100 scf/ton (<3.1 m^3/t). However, migration and conventional trapping of thermogenic and/or biogenic gases can result in unusually high gas contents in low-rank coals.

Over 50 vitrinite reflectance (VR) values from 10 Mesaverde wells in the study area were obtained from coalbed methane operators, Law (1984), and MacGowan and Britton (1992) and used in constructing a Mesaverde coal-rank map. Unfortunately, with the exception of one sample from Law (1984), all of the measured VR data are restricted to the eastern half of the basin. Proximate and ultimate data from 39 samples along the eastern margin of the basin were used to supplement the measured vitrinite reflectance data. The heating value of the coal (Btu/lb) was calculated on a mineral matter moisture free (mmmf) basis and then converted to equivalent vitrinite reflectance values using the polynomial equation

$$VR = 0.87302 - (1.35 \times 10^{-4})(Btu) + (9.14 \times 10^{-9})(Btu)^2. \quad (1)$$

This equation (fig. 27a), determined by regression line using coal-rank data from Murray and others (1977), Stach and others, (1982) and American Society for Testing Materials (1983), can be used to estimate vitrinite reflectance values for high-volatile A bituminous and lower rank coals. A comparison of measured vitrinite reflectance values with vitrinite reflectance data calculated from equation (1), using coal rank data provided by the operators and from Tremain and Toomey (1983), shows that calculated vitrinite reflectance values less than 0.78 percent (high-volatile B bituminous and lower rank) generally fall within 0.1 percent VR of the measured values (fig. 27b). However, the calculated vitrinite reflectance values tend to be slightly overestimated in the high-volatile A bituminous range. In the absence of measured vitrinite reflectance data and proximate and ultimate analyses, vitrinite reflectance profiles of Mesaverde coals and shales were used. Vitrinite reflectance profiles were constructed from data obtained from coalbed methane operators, Tremain and Toomey (1983), Law (1984), and MacGowan and Britton (1992). Regression analyses were performed on the vitrinite

***In* Kaiser, W. R., and others, 1993, Geologic and hydrologic controls on coalbed methane: Sand Wash Basin: The University of Texas at Austin, Bureau of Economic Geology, topical report prepared for the Gas Research Institute under contract no. 5091-214-2261 (GRI–92/0420), p. 51-62.**

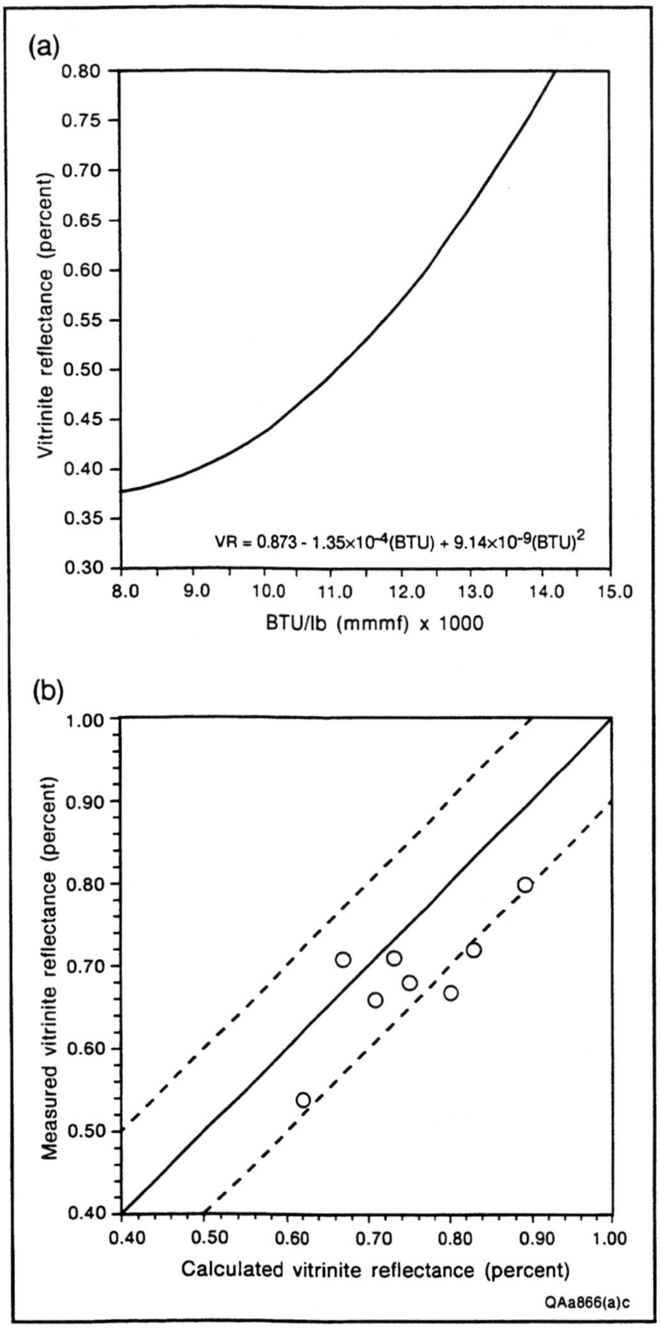

Figure 27. Correlation between vitrinite reflectance and coal Btu (mmmf). (a) Equation and graph used to convert Btu data into equivalent vitrinite reflectance values. (b) Comparison between calculated and measured vitrinite reflectance values. Vitrinite reflectance, proximate, and ultimate data were provided by operators and by Tremain and Toomey (1981).

reflectance data from profiles in the Sand Wash and Washakie Basins (fig. 28). The equations calculated by regression analyses (fig. 28) were subsequently used to estimate vitrinite reflectance values for the top of the Mesaverde at approximately 55 locations in the Washakie and Sand Wash Basins (fig. 28). Vitrinite reflectance versus depth profiles for Mesaverde coals in the Sand Wash Basin were also generated to estimate the amount of overburden removed but were not used in calculating coal rank (fig. 29). The logarithmic increase in vitrinite reflectance values with increasing burial depth (figs. 28 and 29) is common in many western basins (Tyler and others, 1991). Although vitrinite reflectance profiles are useful for estimating coal rank and gas generating stages in a basin, estimated vitrinite reflectance values alone cannot be used to interpret burial history and the timing of structural element formation. Calculated vitrinite reflectance values will be underestimated if uplift and subsequent removal of section from areas of higher-rank coals has occurred. Furthermore, calculated values can overestimate or underestimate coal rank if different parts of the basin have had significantly different paleogeothermal gradients and/or coalification histories. However, a generalized interpretation of the burial history is possible using additional information from other sources.

Coal rank ranges from high-volatile C bituminous along the eastern and southern margins of the basin to medium-volatile bituminous in the Sand Wash Basin's structural center along the Little Snake River and up to the semianthracite rank in the Washakie Basin (fig. 30). Coal rank in the eastern part of the basin, where the thickest coals are located, is generally high-volatile C to B bituminous rank, although some basinward coal beds may approach high-volatile A bituminous rank. The lower coal ranks along the basin margins, White River Uplift, and the eastern part of the Cherokee Arch suggest that these structures probably started to form during the Paleocene and Eocene (Tremain and others, this vol.; Johnson and Nuccio, 1986) before the main stage of coalification. Previous studies on the timing of coalification in the Piceance and San Juan Basins suggest that maximum temperatures and burial depths were attained during the Late Eocene and Oligocene (Johnson and Nuccio, 1986; Law, 1992). Tertiary intrusives in the Elkhead Mountains were emplaced during the Oligocene to Miocene, and upper Tertiary dikes were formed during the middle to late Miocene (Tweto, 1979; Tyler and Tremain, this vol., fig. 7). Therefore, maximum burial depth and coalification in the Sand Wash Basin may also have occurred during this time, although the exact timing of maximum burial and coalification remains uncertain.

Assuming that vitrinite reflectance values are approximately 0.2 to 0.3 percent at the surface (Teichmüller and Teichmüller, 1981), the relatively high vitrinite reflectance values of 0.4 to 0.5 percent along the basin margins suggest that the basin has probably undergone significant uplift and erosion following the main stage of coalification. The amount of overburden removal can be approximated using the equations

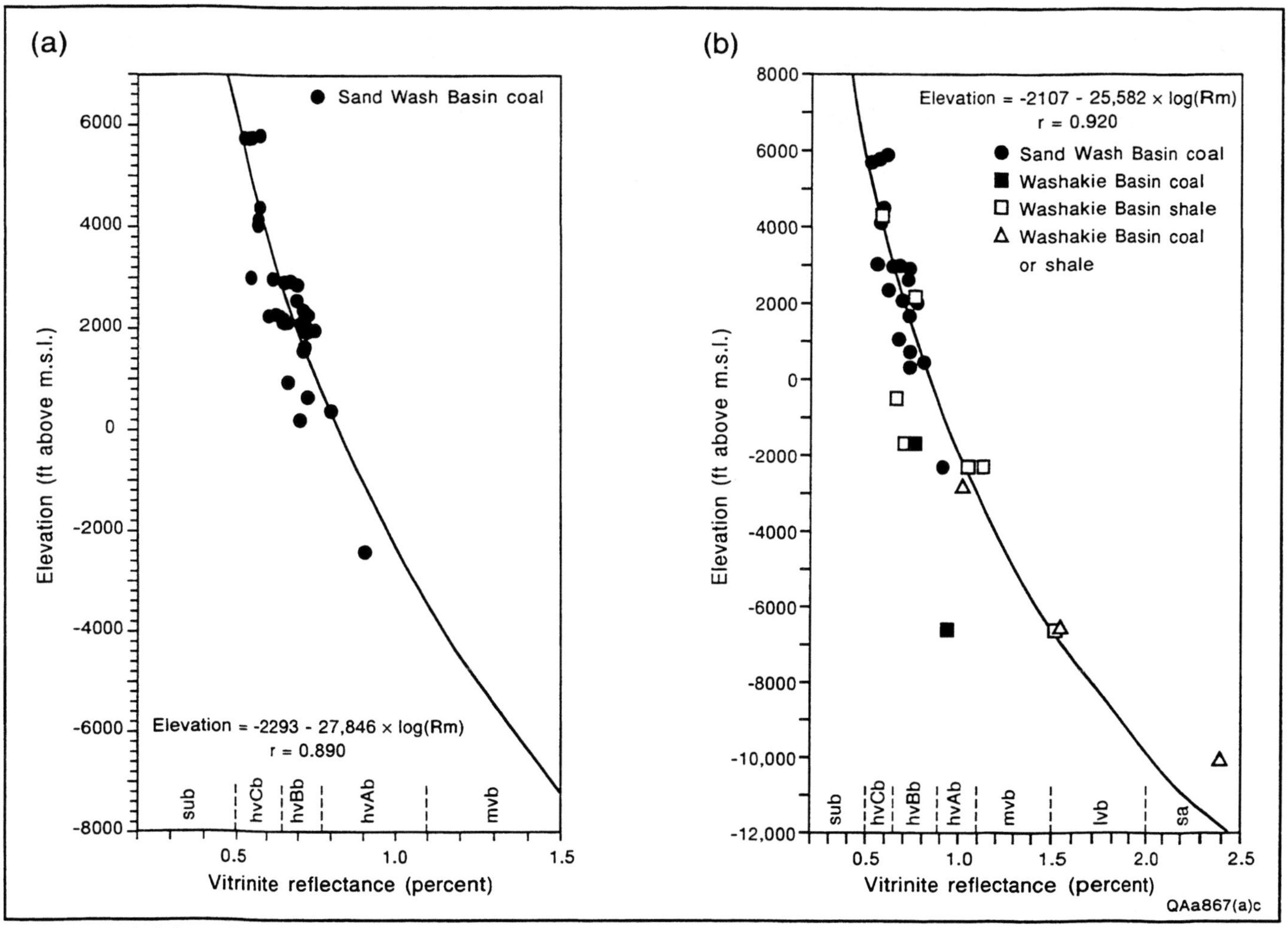

Figure 28. Vitrinite reflectance profiles, Sand Wash Basin. (a) Elevation versus vitrinite reflectance values of Mesaverde coals in the Sand Wash Basin. (b) Elevation versus vitrinite reflectance values of Mesaverde coals and shales from the Sand Wash and Washakie Basins. The equations in (a) and (b) were used to estimate vitrinite reflectance values based on depth in the two basins.

determined from vitrinite reflectance profiles in figure 29. Both equations in figure 29 have essentially the same correlation coefficient, but the amount of overburden removal estimated from each equation differs significantly depending on which equation and which surface vitrinite reflectance value are used. Overburden removal estimates range from 2,600 ft (793 m) (fig. 29; surface VR = 0.3 percent) to 10,300 ft (3,139 m) (fig. 29; surface VR = 0.2 percent). However, assuming a surface vitrinite reflectance value of 0.25 percent, the amount of overburden removed from the basin ranges between 3,500 and 7,600 ft (1,067 and 2,316 m). This range of overburden removal is probably a more reasonable estimate and is similar to the amount of overburden removed in the Piceance Basin (Johnson and Nuccio, 1986).

A major northwest-trending fault system extending northwest from Craig cuts the Miocene Browns Park Formation (Tweto, 1979), suggesting that the main stage of coalification, during which maximum gas generation was attained (Late Paleocene and Oligocene), may have occurred before the fault system formed. However, this fault system could also have developed simultaneously with the main stage of thermogenic gas generation and been reactivated during regional uplift in the late Miocene, thus cutting the Brown's Park Formation. Therefore, it is not known at this time whether or not this fault system was in place to conventionally trap migrating thermogenic gases. However, conventional trapping of thermogenic and/or biogenic gases after the present-day hydrologic regime developed has probably occurred. Active hydrocarbon overpressure may be present in the deeper parts of the Sand Wash Basin (Scott and Kaiser, this vol.). Coals in the Washakie Basin are buried much deeper and have reached the semianthracite rank, suggesting that active hydrocarbon overpressure is active (McPeek, 1981).

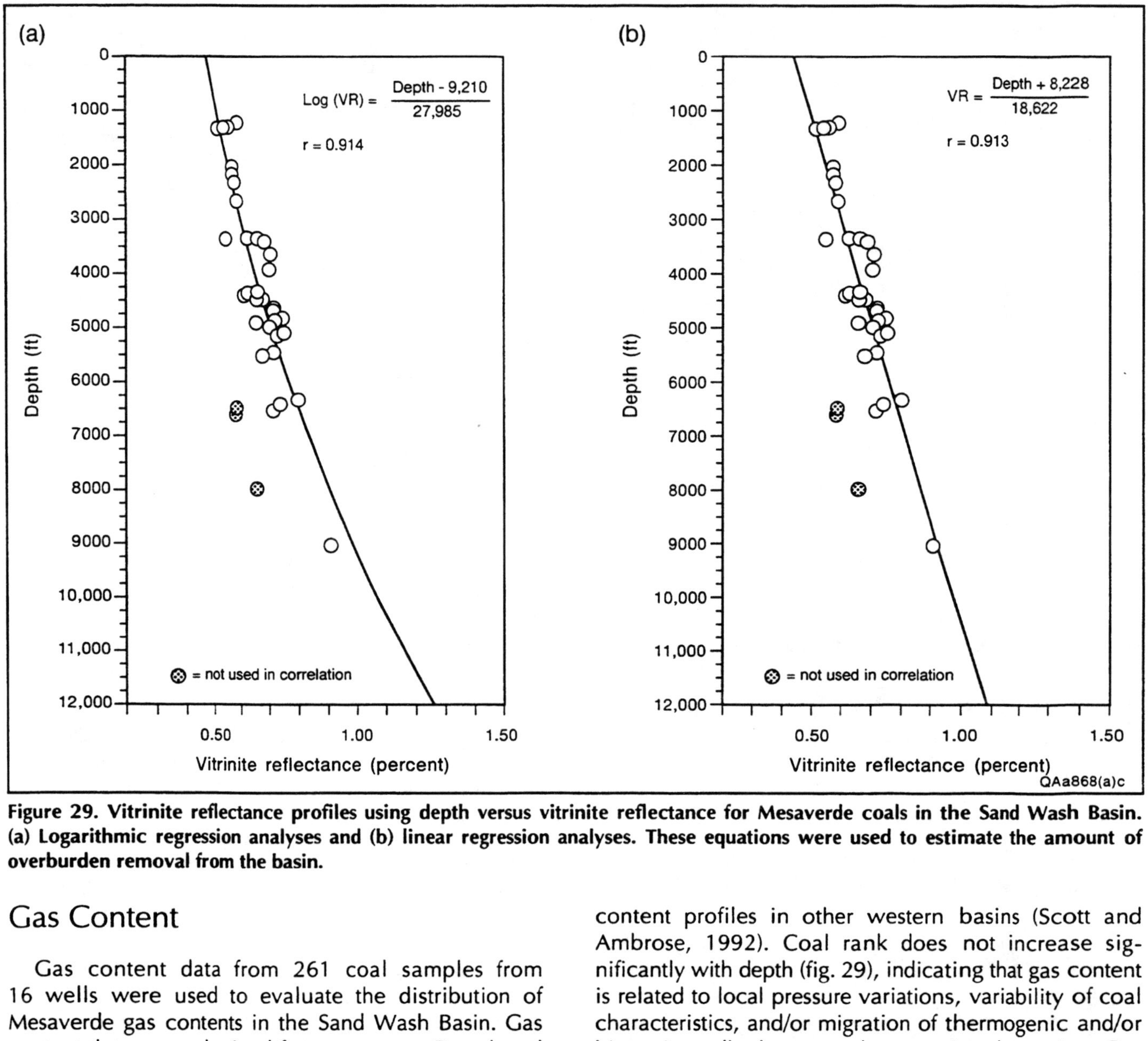

Figure 29. Vitrinite reflectance profiles using depth versus vitrinite reflectance for Mesaverde coals in the Sand Wash Basin. (a) Logarithmic regression analyses and (b) linear regression analyses. These equations were used to estimate the amount of overburden removal from the basin.

Gas Content

Gas content data from 261 coal samples from 16 wells were used to evaluate the distribution of Mesaverde gas contents in the Sand Wash Basin. Gas content data were obtained from operators, Boreck and others (1983), and Tremain and Toomey (1983). All gas content readings were measured by the U.S. Bureau of Mines method and were corrected to an ash-free basis when proximate data were available; ash content ranges from less than 1 to 28.2 percent and averages 9.2 percent. In the absence of proximate data, all ash content values from the same well were averaged in order to correct the gas contents to a calculated ash-free basis. Mesaverde gas content (ash-free) data for all samples range from less than 1 to more than 540 scf/ton (<0.1 to >16.9 m^3/t) but are generally less than 200 scf/ton (<6.3 m^3/t) (average 147 scf/ton [4.6 m^3/t]) (fig. 31).

Gas content versus depth profiles show a gradual increase in gas content and wide scatter of gas content data with increasing burial depth (fig. 32) similar to gas content profiles in other western basins (Scott and Ambrose, 1992). Coal rank does not increase significantly with depth (fig. 29), indicating that gas content is related to local pressure variations, variability of coal characteristics, and/or migration of thermogenic and/or biogenic coalbed gases and conventional trapping. Gas contents are less than 20 scf/ton (<0.6 m^3/t) for samples shallower than 1,000 ft (<305 m), indicating that coalbed gases may have migrated out of the system due to low confining pressures and/or lack of seals. Factors controlling gas content measurements include sample type, sampling procedures, coal properties, and analytical methods and quality. Scatter of gas content data probably reflects experimental and handling procedures and/or the type of sample. Gas content measurements for core samples are significantly greater than gas content values for cuttings and sidewall core samples. Comparison of gas content values of whole core samples with sidewall core and cutting samples over approximately the same depth interval indicates that whole core gas content measurements are 1.6 and 1.4 times greater than sidewall core and cutting samples, respectively. However, gas

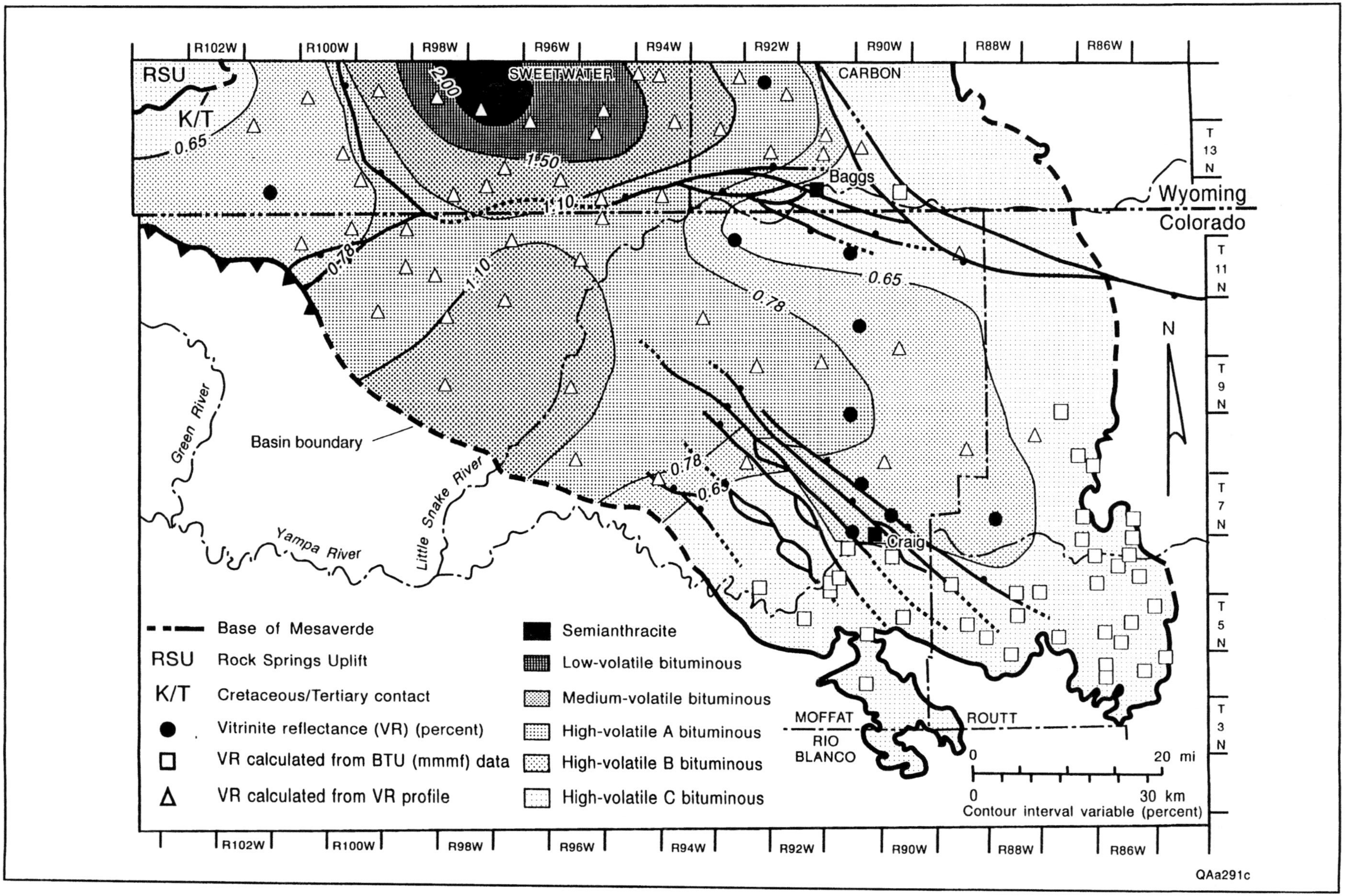

Figure 30. Mesaverde coal rank map. Coal rank is mainly high-volatile C to B bituminous in the eastern part of the basin and reaches medium volatile bituminous in the basin's structural center. Measured vitrinite reflectance values are from operators, Tremain and Toomey (1982), Law (1984), and MacGowan and Britton (1992).

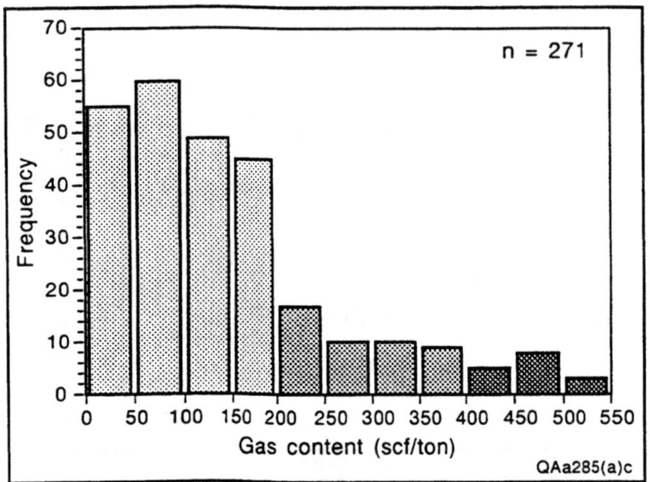

Figure 31. Histogram of ash-free Mesaverde gas contents. Most gas contents are less than 200 scf/ton (6.3 m³/t) but locally exceed 400 scf/ton (12.5 m³/t) in some coal beds.

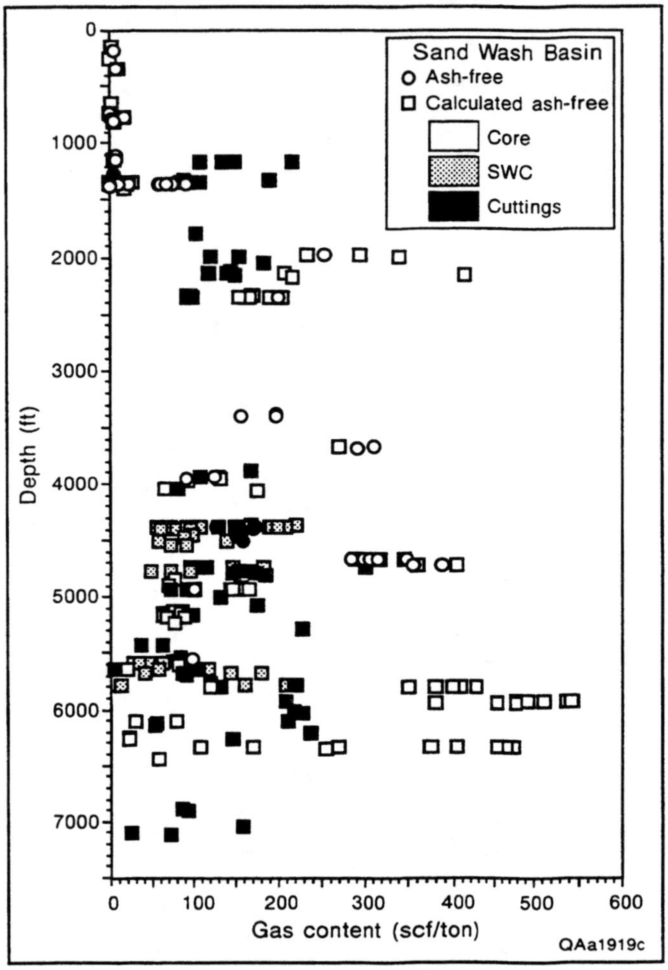

Figure 32. Gas content profile for the Sand Wash Basin Mesaverde coals. Gas content increases with depth, as in other western coal basins. At any particular depth, there is a wide range of gas content values. Sidewall cores and cuttings generally have lower gas contents than whole cores, indicating that sampling and handling procedures influence gas content measurements. Data are ash-free values calculated from average ash contents of adjacent seams.

contents from whole core samples within several feet of each other can show a large difference in gas content values (fig. 32), indicating that factors other than sample type affect gas content values.

Most gas content measurements are performed at room temperature rather than reservoir temperature. Since gas is desorbed more rapidly from coal surfaces at higher temperatures, gas contents measured at reservoir temperatures are usually higher than gas content measurements taken at room temperature (fig. 33). Gas contents determined at reservoir temperatures (98° to 130°F [37° to 54°C]) are generally 1.2 times higher than gas contents determined at room temperature. However, some gas content measurements made at 130°F (54°C) are significantly higher than gas contents made at room temperature (fig. 33). Factors behind this variability in gas content values between room temperature and higher reservoir temperatures are uncertain at this time but may be related to methane diffusivity in the coal. Coals with low diffusion coefficients will retain more methane than coals with high diffusion coefficients. Therefore, coals with low diffusion coefficients may have unusually high gas contents at reservoir temperatures simply because they contain more residual methane than coals with high diffusion coefficients. However, although gas content readings were performed at reservoir temperatures, these experiments were not carried out at reservoir pressures, suggesting that the additional methane released at higher temperatures may remain sorbed on the coal at reservoir pressures.

Factors controlling the distribution of gas contents in coal beds include coal rank, the presence or absence of seals, stratigraphic or structural traps, coal characteristics, local pressure variations, and basin hydrodynamics. Gas content measurements of coal beds in the Sand Wash Basin show a gradational increase in gas content with increasing burial depth and pressure. However, the gas contents of several wells are higher than gas contents in other wells over equivalent depth intervals. The Morgan Federal 12-12 (T8N, R93W, Sec. 12) and Van Dorn No. 1 (T7N, R90W, Sec. 29) are on the downthrown side of a major northwest-trending fault system extending from near Craig (T6N, R90W), 30 mi (48 km) northwestward to T10N, R94W (Tyler and Tremain, this vol., fig. 5). Maximum gas contents in these wells range from more than 300 to 500 scf/ton (>9.4 to 15.7 m³/t) between 4,500 and 6,500 ft (1,372 and 1,981 m). Other Mesaverde coals with anomalously high gas contents are located east of Baggs (T12N, R90W) in an area of artesian overpressuring (Scott and Kaiser, this vol., figs. 32 and 34) along the eastern part of the Cherokee

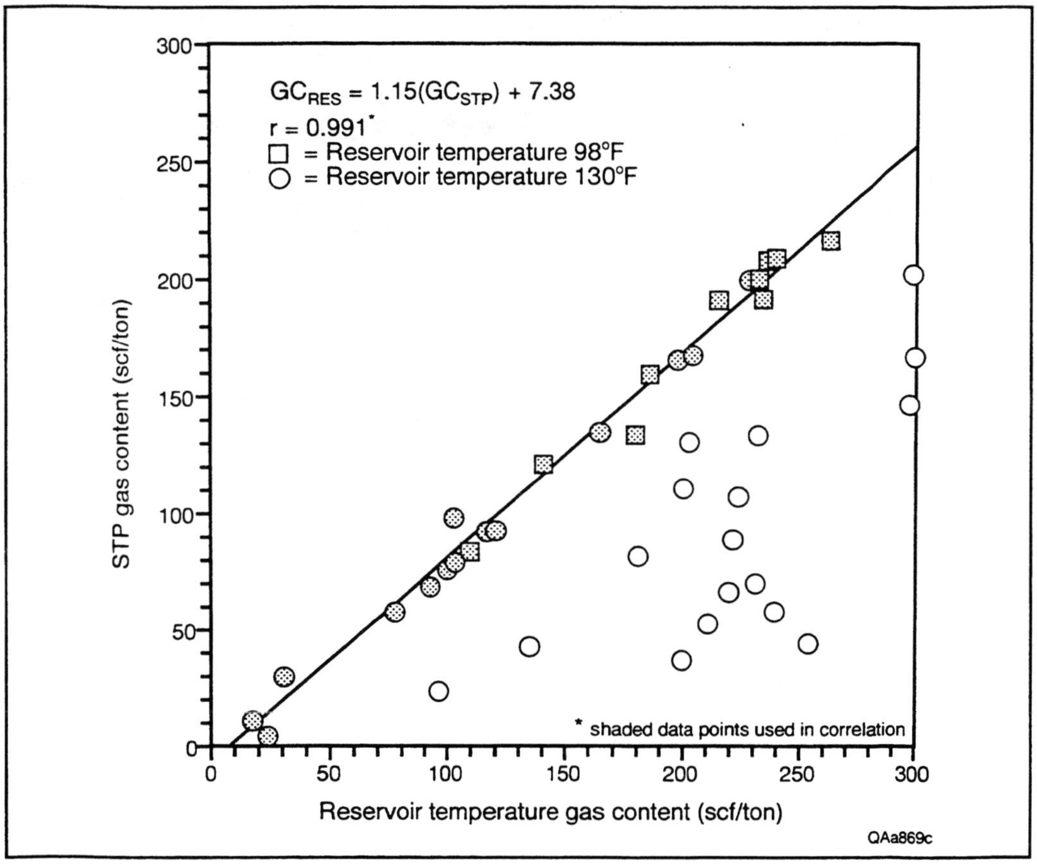

Figure 33. Relation between gas contents determined at room (STP conditions) and reservoir temperatures.

Arch. Maximum gas contents for these wells range from more than 170 to 300 scf/ton (>5.3 to 9.4 m³/t) over a depth interval of 1,000 to 2,400 ft (305 to 732 m).

Gas contents change vertically between coal beds and laterally within individual coal beds between wells (fig. 34). The variability in gas content values could be due to variations in pressure between seams, sample type, coal characteristics, analytical methods and quality, and/or migration of gases in coal beds. Anomalously high Mesaverde gas contents adjacent to the major northwest-trending fault system and along the eastern portion of the Cherokee Arch may be due to migration and conventional trapping of biogenic and/or thermogenic coalbed gases, as well as overpressured conditions. Non-ash-free gas content for coals at 5,900 ft (1,798 m) in the Morgan Federal 12-12 (T8N, R93W, Sec. 12) average 414 scf/ton (13.0 m³/t) (fig. 34). These coals pinch out behind a northeast-trending shoreline sandstone (fig. 34; Hamilton, this vol., figs. 15, 16, and 18). Furthermore, this well is also located on the downthrown (northeast) side of a northwest-trending fault system (Tyler and Tremain, this vol., fig. 7), suggesting that the high gas contents may be due to a combination structural and stratigraphic trapping of migrating gases. Migrating gases could have been trapped during the main stage of coalification, depending on the timing of fault development, and/or during migration of early thermogenic and/or biogenic gases transported basinward by ground water.

Sorption Isotherms

Adsorption analyses of eight Mesaverde coal samples from three wells were available for isotherm evaluation (fig. 35). All isotherms were converted to an ash-free basis. Adsorption isotherms for the Van Dorn well are similar to some Fruitland coal isotherms from the same rank in the San Juan Basin. Significantly more gas can be adsorbed on coals from the Van Dorn No. 1 well (T7N, R90W, Sec. 29) than on coals from the Colorado State No. 1-31 and Blue Gravel wells (T7N, R88W, Sec. 31 and T8N, R91W, Sec. 3, respectively). Gas-content estimates from all of these wells fall below the adsorption isotherms, indicating that the coals are undersaturated with respect to methane. Coal beds in the Morgan Federal 12-12 No. 33-3 well (T8N, R93W, Sec. 12) have estimated gas contents in excess of 300 to 500 scf/ton (>9.4 to 15.7 m³/t) but do not produce methane immediately upon completion. This indicates

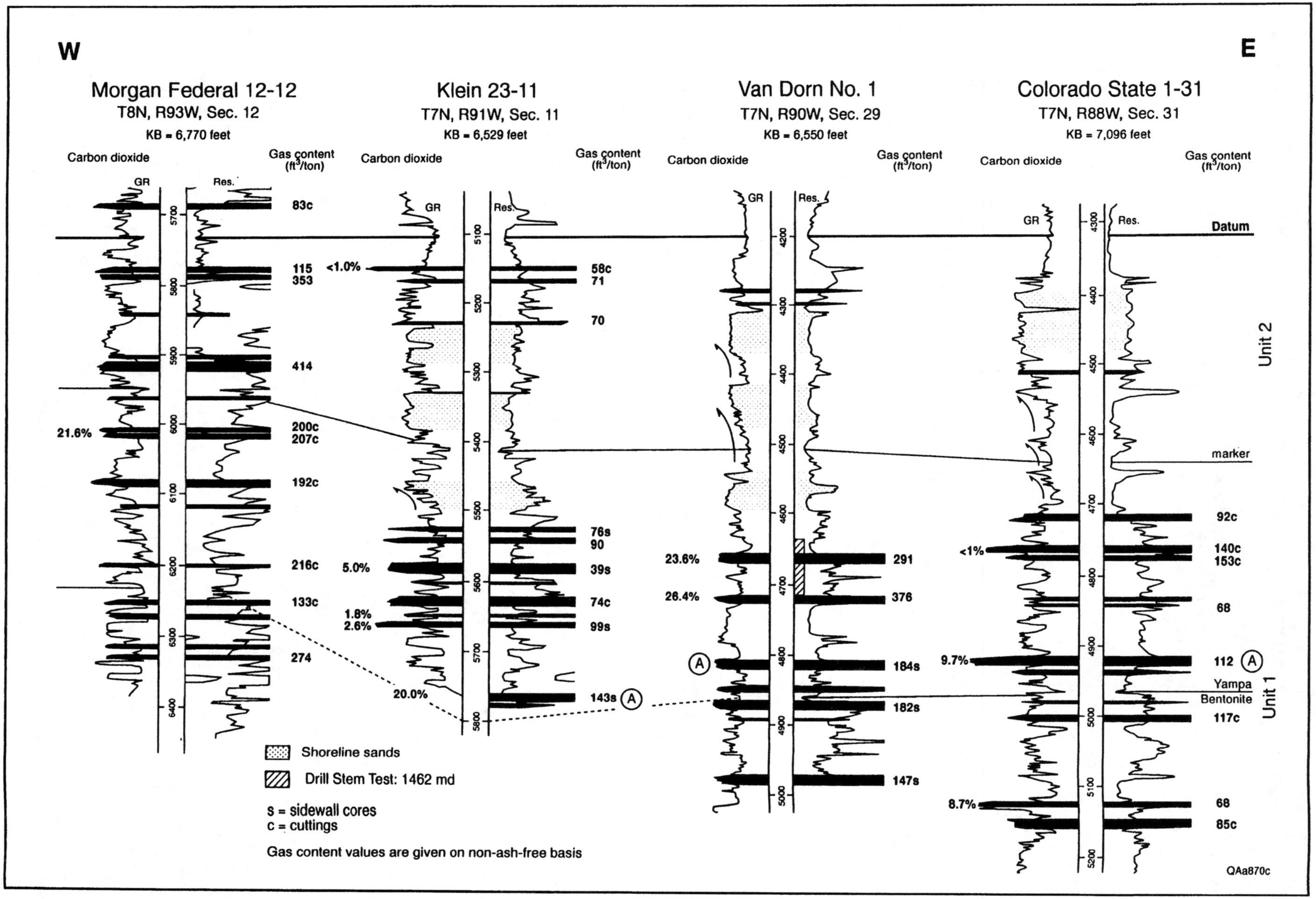

Figure 34. West-east cross section showing the changes in gas content and gas composition between different Mesaverde coal beds. The high gas contents in coal beds at 5,900 ft (1,799 m) in the Morgan Federal 12-12 well may be due to stratigraphic and structural trapping of migrating coalbed gases.

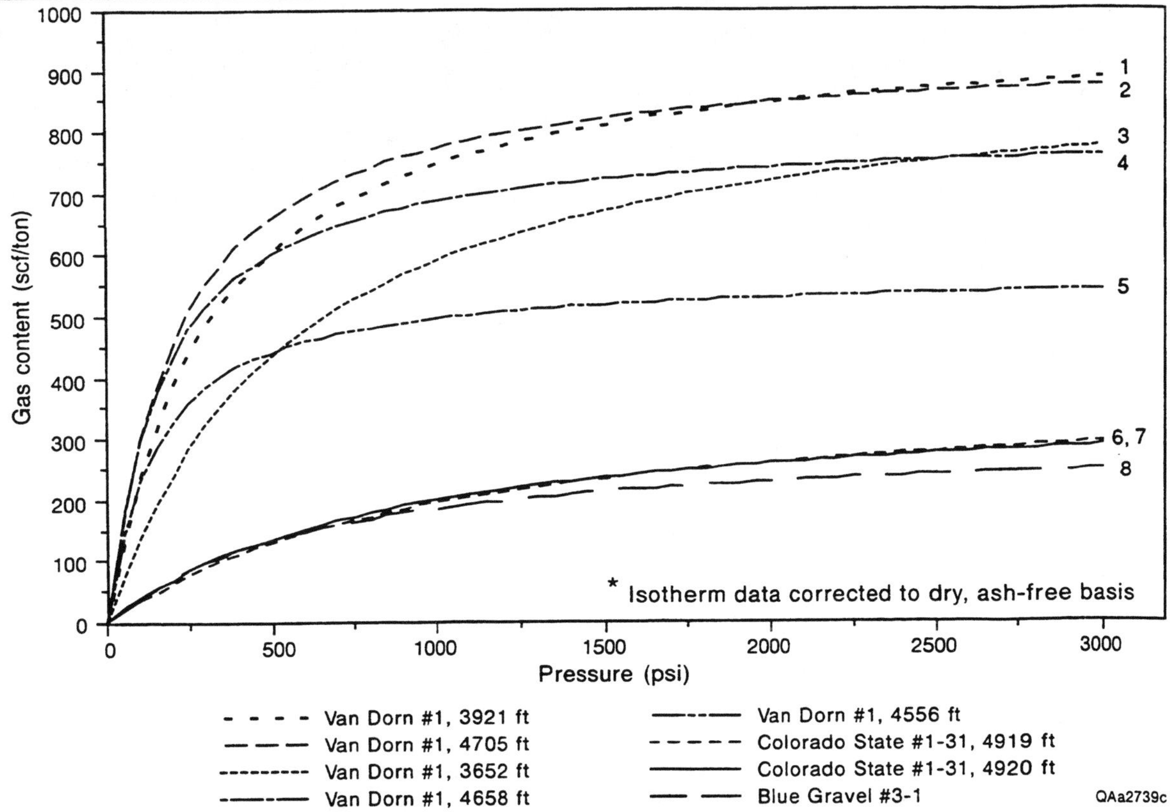

Figure 35. Adsorption isotherms for Mesaverde coalbed gases. The variability in the adsorptive capacity of coal may explain the wide range of gas contents of Mesaverde coals.

that these coals are also undersaturated with respect to methane and that isotherms for these coals are probably similar to isotherms from the Van Dorn No. 1 well (fig. 35).

The reasons adsorption isotherms in the Colorado State No. 1-31 and Blue Gravel No. 3-1 are significantly lower than isotherms in the Van Dorn No. 1 well remain unclear. Isotherm variability is not obviously related to coal rank because all the samples are from essentially the same rank coal (high-volatile A to B bituminous). However, subtle changes in the coal surface structure related to the generation, retention, and/or migration of volatile matter during thermal maturity can affect the sorption capacity of a coal. Major changes in coal surface structure occur during the first coalification jump when the coal enters the oil-generating stage (vitrinite reflectance values of approximately 0.5 percent). These changes include (1) continued moisture loss and decrease in oxygen content of the coal, (2) generation of wet gases and heavier hydrocarbons, and (3) pore-size distribution. The decrease in methane sorption with increasing moisture content is directly related to the amount of oxygen in the coal (Joubert and others, 1973, 1974). The oxygen content for curves 1 and 2 (fig. 35) is approximately 10 percent, whereas the oxygen content for curves 3 through 5 (fig. 35) averages over 11 percent, suggesting that a possible increase in inherent moisture may partially explain adsorption isotherm variability in coals from the Van Dorn No. 1 well. Unfortunately, ultimate analyses were not available for coals from the Colorado State No. 1-31 and Blue Gravel No. 3-1 wells. The movement of ground water through coal may also result in coal oxidation and subsequent changes in coal surface properties and adsorptive capacity. However, the effects of ground-water oxidation on coal surface properties have not been extensively evaluated.

The methane sorption capacity of a coal is also related to the distribution of less than 12A pores in the coal and can differ significantly between coal having slightly different carbon contents (Schwarzer, 1983). However, there was no correlation between carbon content and the adsorption isotherms in Mesaverde coals. The assumption that methane sorption capacity increases continually with increasing coal rank is not valid (Schwarzer, 1983). Coal surface area and sorption capacity decrease rapidly over vitrinite reflectance values between between 0.6 and 0.8 percent (Ettinger and others, 1966; Thomas and Damberger, 1976). Both surface area and sorption capacity progressively increase during the medium-volatile bituminous and higher coal ranks (Moffat and Weale, 1955; Ettinger and others, 1966; Thomas and Damberger, 1976). This decrease in surface area and methane sorption capacity corresponds to the wet-gas-generating stage of the coal beds. The generation of wet gases and other hydrocarbons from the coal plugs micropores and limits methane accessibility. Coal surface area decreases from approximately 250 to 50 m^2/g over the vitrinite reflectance range of 0.6 to 0.8 percent (Thomas and Damberger, 1976). Therefore, high-volatile bituminous coals have the potential to show large variability in gas sorption capacities during relatively small changes in coal rank due to subtle changes in coal structure and the distribution of micropores.

Gas Composition

The composition of coalbed gases is directly related to coal rank, basin hydrodynamics, and maceral composition (Scott and Kaiser, 1991). The gas dryness index (the ratio of methane to methane through pentane; C_1/C_{1-5}) reflects the amount of chemically wet gases generated during the thermal maturation of hydrogen-rich coals. In general, hydrogen-rich coals in the oil-window or oil-generating stage (vitrinite reflectance of 0.5 to 1.2 percent) produce significant amounts of wet gases (ethane, propane, etc.), whereas coals having vitrinite reflectance values less than 0.5 percent or greater than 1.2 percent will generate relatively few wet gas components and have C_1/C_{1-5} values near unity (Scott and others, 1991a). The chemistry of coalbed gases can be significantly altered through biogenic activity. Bacterial alteration of chemically wet gases can remove nearly all of the wet gas components, producing chemically dry gases resembling thermogenic methane (James and Burns, 1984). Furthermore, mixtures of biogenic and thermogenic coalbed gases are difficult to recognize using only gas dryness indices and methane isotopic data. The isotopic composition of carbon dioxide from coal beds may prove to be more useful in determining the biogenic or thermogenic nature of coal bed gases than methane isotopic data alone, particularly when mixtures of thermogenic and biogenic methane may be present.

The chemical compositions of desorbed gas samples from 36 coal samples in 6 Mesaverde wells were used to evaluate the chemical composition and origin of Williams Fork coalbed gases. Although no produced coalbed gases in the basin were available for analysis, the compositional ranges of a large number of desorbed coalbed gases will approximate the compositional ranges of produced gases (Scott, 1993). Desorbed coalbed gases will generally contain more carbon dioxide, nitrogen, and wet gas components (Mavor and others, 1991; Scott, 1993), particularly if higher temperatures are used during desorption. The gas dryness index ranges from 0.79 to 1.00 and averages 0.95 (fig. 36). These values are similar to Fruitland coalbed gases in the San Juan Basin (C_1/C_{1-5} range of 0.77 to 1.00; average of 0.96; Scott and others, 1991a, b). Carbon dioxide content for Mesaverde coal beds ranges from less than 1 to more than 25 percent (fig. 34). The range of carbon dioxide content for Mesaverde coalbed gases is also similar to Fruitland coalbed gases, which range from less than 1 to more than 25 percent (Scott and others, 1991a, b; Scott, 1993). The average carbon dioxide content of Mesaverde coals in the Sand Wash Basin (6.7 percent) is similar to the average carbon dioxide content of coals from the northern part of the San Juan Basin (6.4 percent; Scott and others, 1991a, b) and slightly more than the overall average of Fruitland carbon dioxide content (4.5 percent). Nitrogen content in Mesaverde coalbed gases ranges from less than 1 to 20 percent and averages approximately 4 percent. This average nitrogen content is significantly higher than the average Fruitland coalbed nitrogen values (<0.1 percent; Scott and others, 1991a, b). The higher average nitrogen values of Williams Fork coalbed gases may be due to gas sampling; these gases were desorbed from coal samples, which increases the possibility of air contamination, whereas Fruitland data are from produced coalbed gases.

Gas composition changes vertically between coal beds within individual wells and laterally between wells (fig. 34). However, at least one coal bed (fig. 34), which can be traced laterally over several tens of miles using density log profiles, has consistently high carbon dioxide values (fig. 34) near or above 10 percent. This suggests that factors controlling gas composition such as basin hydrodynamics, gas migration, maceral composition, biogenic activity, or a combination of these factors, can operate consistently over laterally extensive areas in continuous seams. Coals with high carbon dioxide contents are generally characterized by high C_1/C_{1-5} values (fig. 37). Furthermore, coal beds in the lower part of the Williams Fork Formation (Units 1 and 2) contain more carbon dioxide and fewer wet gas components than coals in Units 3 and 4 of the upper Williams Fork Formation. However, coal beds from the Morgan Federal 12-12 (T8N, R93W, Sec. 12) tend to have chemically wet gases and relatively high carbon dioxide content (fig. 34).

Origin of Coalbed Gases

Early thermogenic, thermogenic, and secondary biogenic gases are found in coal beds (Scott, 1993). Early thermogenic gases are formed between vitrinite reflectance values of 0.5 and 0.8 percent, whereas thermogenic gases are generated after the threshold of methane generation has been reached. Primary biogenic gases, generated during the early stages of coalification, are probably not preserved in coal beds (Scott, 1993). Secondary biogenic gases (Scott, 1993) are formed through bacterial degradation of chemically wet coalbed

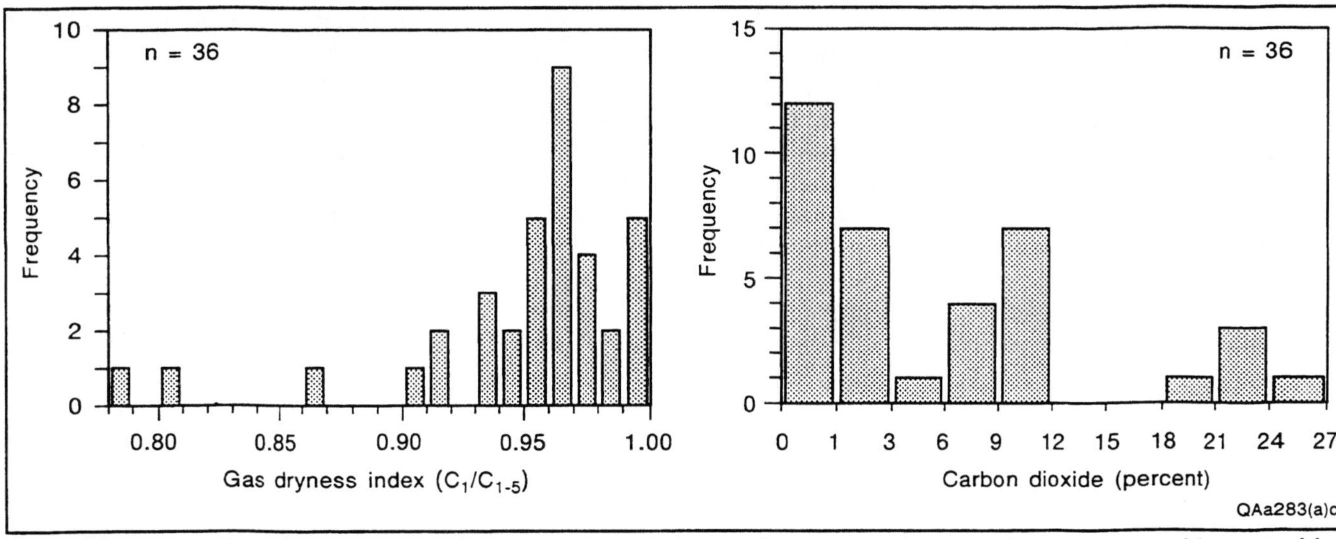

Figure 36. Composition of Mesaverde coalbed gases. Desorbed gases have a wide range of chemical compositions. Coal beds have entered the early gas generation stage as indicated by the minor amounts of wet gases in the samples. High carbon-dioxide contents in some coal beds may reflect bacterial activity, gas migration, and/or variations in maceral composition.

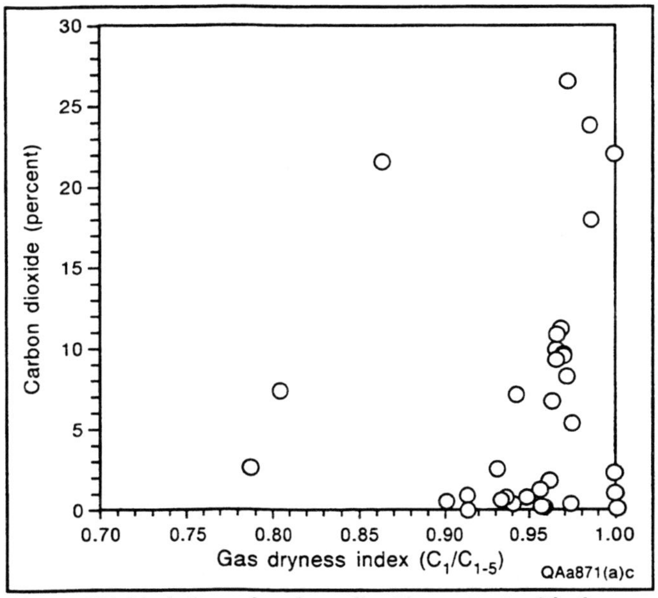

Figure 37. Variation of carbon dioxide content with the gas dryness index (C_1/C_{1-5} values). Coals with the higher carbon dioxide content generally have very dry gases, although coals in the Morgan Federal 12-12 and Colorado State 1-31 wells have wet gases.

gases and organic compounds on the coal by bacteria transported in meteoric water flowing basinward from a recharge area (Scott and others, 1991a, b; Kaiser and others, 1991b; Scott and Kaiser, 1991). Determining the source of methane and carbon dioxide in coalbed gases is important for evaluating origin of coalbed gases and the migration of coalbed gases within the basin. Significant amounts of carbon dioxide are released from coals during maturation. Based on the equations and on data presented by Levine (1987), more than 4,186 scf/ton (>131.2 m³/t) (STP; 77°F [25°C], 1 atm) carbon dioxide and 6,040 scf/ton (189.3 m³/t) methane can be released from vitrinitic material over the bituminous to semianthracite coal rank (VR values of 0.5 to 2.0 percent) during coalification. Assuming that water is released in liquid form, carbon dioxide and methane represent approximately 40 and 60 percent (by volume), respectively, of the total volatiles released from the coal. However, assuming that only methane and carbon dioxide are generated, more than 22,900 scf/ton (>717.8 m³/t) of methane can be generated from hydrogen-rich organic material, whereas approximately 8,800 scf/ton (~275.8 m³/t) of methane is generated from coals composed entirely of vitrinite (Levine, 1987). Therefore, even minor amounts of hydrogen-rich organic matter can dramatically increase the amount of methane generated from coal beds, assuming that no wet gas components are generated.

Significant amounts of carbon dioxide are generated during the early stages of coalification before the main stage of thermogenic gas generation; the amount of carbon dioxide decreases with increasing maturation (Juntgen and Karweil, 1966; Hunt, 1979; Creedy, 1988). Thermogenic carbon dioxide generated during coalification can remain sorbed to the coal surface or be dissolved in formation waters and subsequently transported out of the system. An additional source of carbon dioxide is from bacterial activity. Bacteria transported in ground water moving basinward through coal beds can metabolize the chemically wet coalbed gases to produce biogenic carbon dioxide and methane (Scott and Kaiser, 1991). The origin of carbon dioxide in coalbed gases can be determined from the isotopic composition of the carbon dioxide. Carbon dioxide released during coalification will be depleted in $\delta^{13}C$

having $\delta^{13}C$ values of –25 to –15 ‰. Biogenic carbon dioxide is enriched in $\delta^{13}C$ with $\delta^{13}C$ values ranging from –20 to +30 ‰ (Jenden, 1985), depending on the intensity and duration of bacterial activity. Therefore, carbon dioxide with positive $\delta^{13}C$ values is predominantly biogenic whereas $\delta^{13}C$ values less than –15 ‰ are generally thermogenic; mixtures of biogenic and thermogenic gases fall somewhere between. However, carbon dioxide derived from magmatic sources ($\delta^{13}C$ values of –7 to –9 ‰; Jenden, 1985) should also be considered when evaluating gas origin.

Williams Fork coalbed gases were not available for detailed isotopic analyses. Even with isotopic analyses, gas origin may be difficult to determine, but the maturation level coalbed gas generation can be evaluated based on coal rank data. Vitrinite reflectance profiles, using Mesaverde coals, indicate that coal beds in the eastern part of the Sand Wash Basin at depths of 6,000 ft (1,829 m) are just entering the main stage of thermogenic gas generation (fig. 28). The relatively low rank of these coals suggests that Williams Fork coalbed gases are predominantly early thermogenic and biogenic, although migration of thermogenic gases from coals or shales deeper in the basin may have occurred. However, the migration of main stage thermogenic gases from areas of high coal rank is limited because there are few coal beds located in the thermally most mature part of the basin (fig. 30; Hamilton, this vol., figs. 14, 17, 19, and 20). The distribution of C_1/C_{1-5} values around 0.96 and the low coal rank suggest that some of the coalbed gases are predominantly early thermogenic (fig. 36). Early thermogenic gases, which are generated between vitrinite reflectance values of approximately 0.50 and 0.80 percent, will contain significant amounts of wet gas components. The wettest gases (C_1/C_{1-5} values <0.90) are associated with coals in the main stage of thermogenic gas generation between VR values of 0.5 to 1.3 percent (Scott, 1993). Biogenic gases and coalbed gases from higher-rank coals (vitrinite reflectance values ≥2.0 percent) are composed primarily of methane having C_1/C_{1-5} values near unity.

Carbon dioxide from Williams Fork coal beds is thermogenic, biogenic, or a mixture of both gas types. Calculations of carbon dioxide generation from coal beds using data from Levine (1987) indicate that over 50 percent of the total amount of carbon dioxide generated from vitrinite (type III organic matter) is produced before coals reach the high-volatile A bituminous rank and approximately 17 percent of the total is generated during the high-volatile A bituminous stage. Therefore, a significant portion of the carbon dioxide reported in some Williams Fork coals may have been generated during the earlier stages (high-volatile C and B bituminous) of coalification. However, the timing of carbon dioxide generation and retention in relation to the changes in adsorptive capacity of the coal during coalification remain uncertain or unknown and, therefore, a biogenic source for some of the carbon dioxide cannot be ruled out. Furthermore, the increase in carbon dioxide content with decreasing C_1/C_{1-5} values (fig. 36) suggest that some of the gases may be bacterially derived. The carbon dioxide content of individual seams ranges from less than 2 to more than 20 percent within the same well (fig. 34). However, carbon dioxide content remains consistently high (~10 percent) in some coal beds, which are correlated over tens of miles (fig. 34; Hamilton, this vol., fig. 12). The changes in carbon dioxide content vertically and laterally could be due to variations in maceral composition, which could affect the types and quantities of gases generated from the coal, bacterial activity, and/or migration of coalbed gases. The presence of wet gases with high carbon dioxide values (fig. 36) in the Morgan Federal 12-12 and Colorado State 1-31 wells may indicate migration of coalbed gases. The carbon dioxide is probably indigenous to the coal beds whereas the wet gas components may have originated from shales and carbonaceous shale adjacent to the coal beds or from the coal beds themselves.

Conclusions

1. Mesaverde coal rank ranges from high-volatile C bituminous along the basin margins to medium-volatile bituminous in Sand Wash Basin and semianthracite in the Washakie Basin. Coals in the eastern half of the basin are generally high-volatile C and B bituminous and are in the early stages of thermogenic gas generation.

2. Ash-free Mesaverde gas contents range from less than 1 to more than 540 scf/ton (<0.1 to >16.9 m^3/t) but are generally less than 200 scf/ton (<6.3 m^3/t). Gas contents change vertically and laterally between wells for continuous coal beds in the basin. Gas contents from conventional cores are approximately 1.5 times higher than gas contents than sidewall and cuttings samples over approximately the same depth interval.

3. Factors controlling gas content distribution include coal rank, coal surface characteristics, localized pressure variations, basin hydrodynamics, and conventional trapping of migrating early thermogenic and secondary biogenic gases.

4. Adsorption isotherms of Mesaverde coals are variable. Coal rank, oxygen and carbon content, coal surface properties, and the distribution of 12OA pores affect adsorption isotherms.

5. C_1/C_{1-5} values range from 0.79 to 1.00 and average 0.96. Carbon dioxide ranges from less than 1 to more than 25 percent. Mesaverde coalbed gases are probably early thermogenic and secondary biogenic.

Hydrologic Setting of the Upper Mesaverde Group, Sand Wash Basin

Andrew R. Scott and W. R. Kaiser

Abstract

The Mesaverde Group is a regional aquifer system of high transmissivity. Coal beds may be the most permeable aquifers. Recharge is received at outcrop over the basin's wet, elevated eastern and southeastern margins. Basinward movement of water from the recharge areas is controlled by permeability, topographic gradient, structural dip, structural grain, and depositional fabric. In the eastern Sand Wash Basin, ground water flows westward for eventual discharge in potentiometric lows basinward and/or toward the Yampa River valley. Hydrocarbon overpressure is present in the deeper parts of the basin, whereas artesian overpressure is present along the eastern part of the Cherokee Arch. Hydrocarbon overpressure and hydropressure are hydraulically separated by some type of permeability barrier, such as fault zones, areas of intense diagenesis, facies changes, or a combination of these. The transition between geopressure and hydropressure is often abrupt, and flow along it is commonly vertically upward. No pressure regime in the hydropressured part of the basin is regionally dominant.

Introduction

In the Sand Wash Basin, the Mesaverde Group is a regional aquifer system, confined below and above by the marine Mancos and Lewis Shales, respectively, regional confining units thousands of feet thick (Hamilton, this vol., fig. 11). Mesaverde hydrology was evaluated in an analysis of hydraulic head, pressure regime, and hydrochemistry in the context of Mesaverde stratigraphy, structure, and depositional setting. To map hydraulic head, equivalent fresh-water heads were calculated from shut-in pressures (SIP) recorded in drill-stem tests (DST) and bottom-hole pressures (BHP) calculated from well head shut-in pressures. Static water levels at the basin's south and east margins were also used to map hydraulic head (Robson and Stewart, 1990). The pressure regime was evaluated on the basis of simple and vertical pressure gradients calculated from screened DST data. Chlorinity and total-dissolved-solids (TDS) maps further defined ground-water-circulation patterns. Mesaverde hydrostratigraphy is reviewed as a prelude to a discussion of Mesaverde hydrodynamics (hydraulic head, pressure regime, and hydrochemistry). In the context of depositional and structural settings, hydrodynamics serves as the basis for interpretation of regional flow.

Hydrostratigraphy

The Williams Fork and Almond Formations are major coal-bearing hydrostratigraphic units in the Mesaverde Group. These units are confined above by the Lewis Shale (Hamilton, this vol., fig. 11) and only partially confined or unconfined below by the Iles Formation, which overlies the marine Mancos Shale. In the eastern part of the basin, regionally extensive marine shales at the base of the Trout Creek and Twentymile Sandstones serve to stratigraphically divide the Williams Fork and Almond into a lower Williams Fork unit and an upper Williams Fork/Almond unit (Hamilton, this vol., figs. 12 and 13). However, these shales may not divide them hydrologically. Hydraulic communication is inferred from similar heads within the Mesaverde in various parts of the basin. For example, heads from the Iles, Williams Fork Units 1 and 4, and Almond in the southern part of the basin differ by 100 ft (30 m) or less. Therefore, those parts of the Mesaverde Group subject to meteoric circulation behave regionally as a hydraulically interconnected aquifer system, or single hydrologic unit.

The Mesaverde is composed of interbedded, permeable coal beds and sandstone of regional and local extent. At the southeast outcrop, coal beds are 10 to 20 times more permeable (50 to 100 md) than the regionally extensive Trout Creek and Twentymile Sandstones (~5 md) (Robson and Stewart, 1990). Drill-stem tests in lower unit coals in the Van Dorn well (T7N, R90W) indicate high permeability (1,462 md; Scott, this vol., fig. 34), whereas relatively lower permeabilities (tens of md) were calculated for upper unit coals. Permeability of coal beds in the Baggs area ranges between 100 and 200 md and averages approximately 170 md. There are no public permeability data for Mesaverde coal beds basinward in the subsurface, although sandstones are reported to have permeabilities of 5 to 100 md (Mountain Fuel Supply Company, 1961; Collentine and others, 1981).

In **Kaiser, W. R., and others, 1993, Geologic and hydrologic controls on coalbed methane: Sand Wash Basin: The University of Texas at Austin, Bureau of Economic Geology, topical report prepared for the Gas Research Institute under contract no. 5091-214-2261 (GRI–92/0420), p. 63-76.**

Hydrodynamics

The hydrodynamics of the Mesaverde Group was established from its potentiometric surface (hydraulic head), formation fluid pressure, and hydrochemistry. Nearly 450 Mesaverde DST data from 176 wells were taken from the Petroleum Information data base, and the actual stratigraphic interval of screened DST data was verified from geophysical logs. The quality of DST data was characterized as (1) good if the final shut-in time was greater than 60 minutes, (2) moderate if the final shut-in time was 30–60 minutes, and (3) unknown if the initial and/or final shut-in times were not reported. Approximately 49 and 34 percent of the data were considered as good and moderate, respectively, whereas 17 percent were of unknown quality. Using the highest pressure recorded, whether initial or final SIP, a pressure gradient (pressure-depth quotient) was calculated for all available DST's. Drill-stem tests with pressure gradients of less than 0.30 psi/ft (<6.8 kPa/m) were eliminated from the data base because of their uncertain validity, reflecting insufficient shut-in time, bad test data, presence of gas, pressure depletion, or a combination of these factors. Furthermore, a plot of elevation versus psi/ft showed a break in the data at approximately 0.30 psi/ft (~6.8 kPa/m) (fig. 38). Hydraulic heads and vertical pressure gradients were calculated from SIP's on a screened data set of 181 Mesaverde DST's from 80 wells. Bottom-hole pressures were converted to pressure heads (BHP/hydrostatic gradient) using a fresh-water hydrostatic gradient of 0.433 psi/ft (9.8 kPa/m) and combined with elevation heads (kelly bushing minus midpoint of test) to obtain equivalent fresh-water heads (fig. 39). More than 155 water analyses from 66 Mesaverde wells were available to evaluate basin hydrodynamics. Chemical analyses used for hydrochemical maps were dominantly of fluids recovered in DST's and secondarily of produced water. The analyses were screened for analytical accuracy, using an ionic balance formula (Edmunds, 1981). In most cases, they balanced exactly, indicating that sodium and/or potassium were determined by analytical difference. Consequently, because of the nature of the fluids analyzed and the exact ionic balance, the water analyses are of questionable validity and were only used to delineate general concentration gradients rather than for detailed contouring of concentrations.

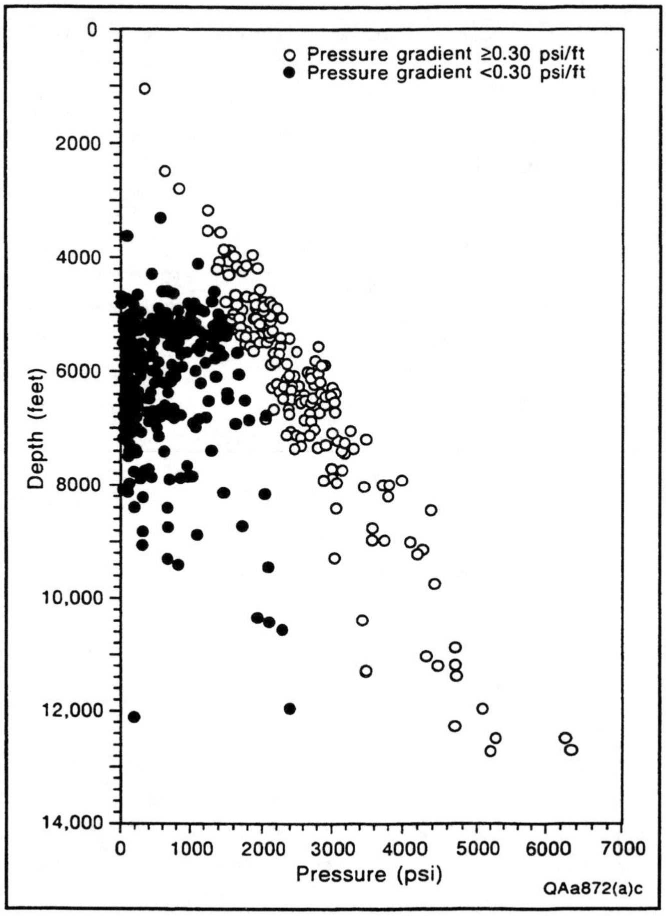

Figure 38. Depth versus pressure plot for Mesaverde DST data from the Sand Wash Basin. DST data with simple pressure gradients less than 0.30 psi/ft (<6.8 kPa/m) were eliminated from the data base because of their uncertain validity.

Potentiometric Surface

An upper Mesaverde Group potentiometric surface map was made from equivalent fresh-water heads using DST data from the Almond Formation and Unit 4 of the upper Williams Fork Formation (Hamilton, this vol., fig. 11). Both stratigraphic units have similar head values, indicating that these two units are in hydraulic communication over much of the basin. However, along the transition boundary between hydrocarbon overpressure and hydropressure, these two units may not be in hydraulic communication. The potentiometric surface map of the upper Mesaverde Group shows potentiometric highs along the topographically higher eastern margin of the basin and along the southern flank of the Rock Springs Uplift (fig. 40), where the higher heads reflect a structural platform (Tyler and Tremain, this vol., fig. 5). From the elevated eastern margin of the basin, the surface slopes toward a potentiometric low in the east-central part of the basin (T10N, R92W). Recharge for the Mesaverde Group occurs along the wet, elevated eastern and southern margins of the basin where annual precipitation exceeds 20 inches/year (>50 cm/yr), whereas recharge over the Rock Springs Uplift is limited by lower precipitation (10 to 12 inches/year [25 to 30 cm/yr]) (fig. 41). Decreasing precipitation

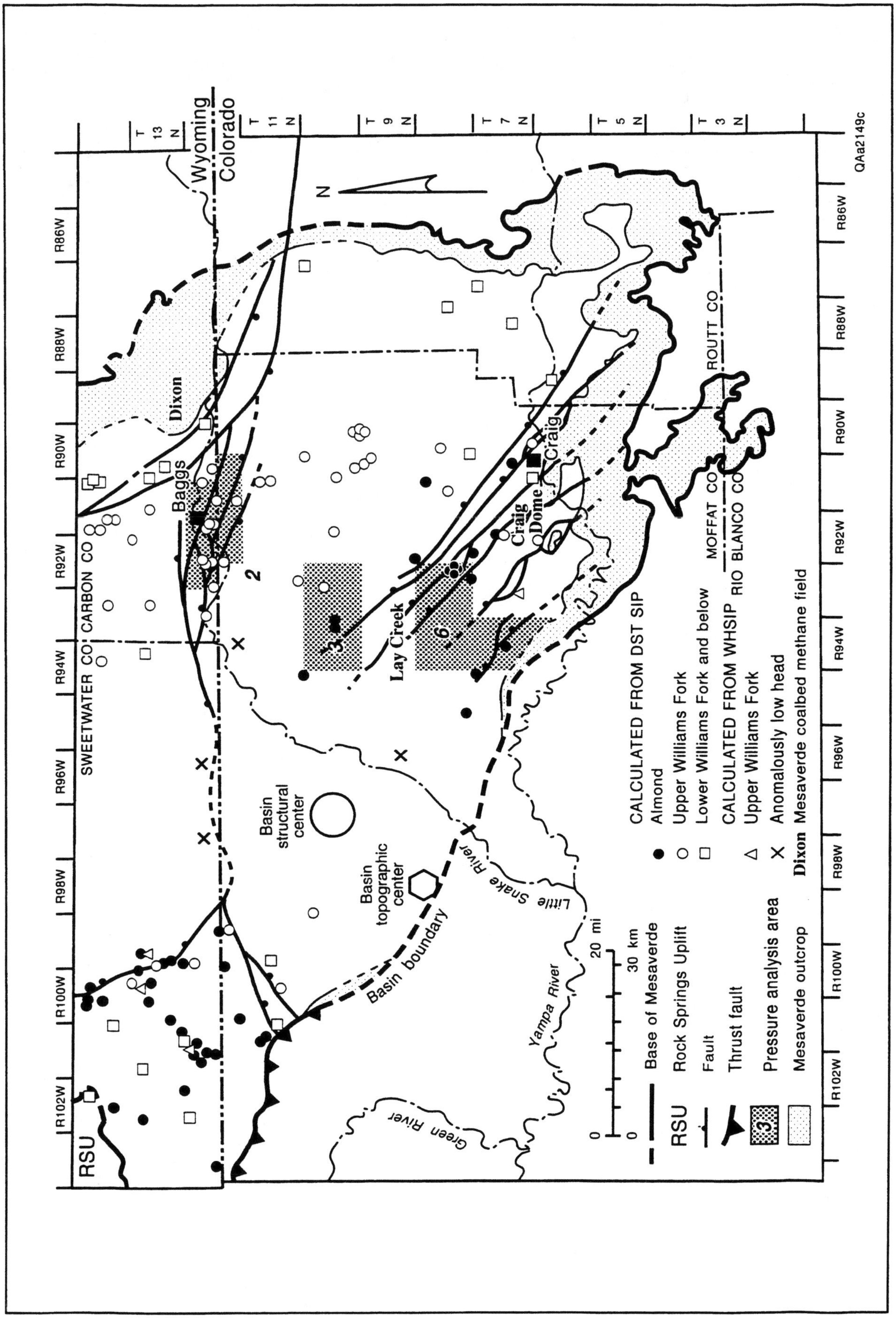

Figure 39. Location of study areas, Mesaverde head data, and coalbed methane fields.

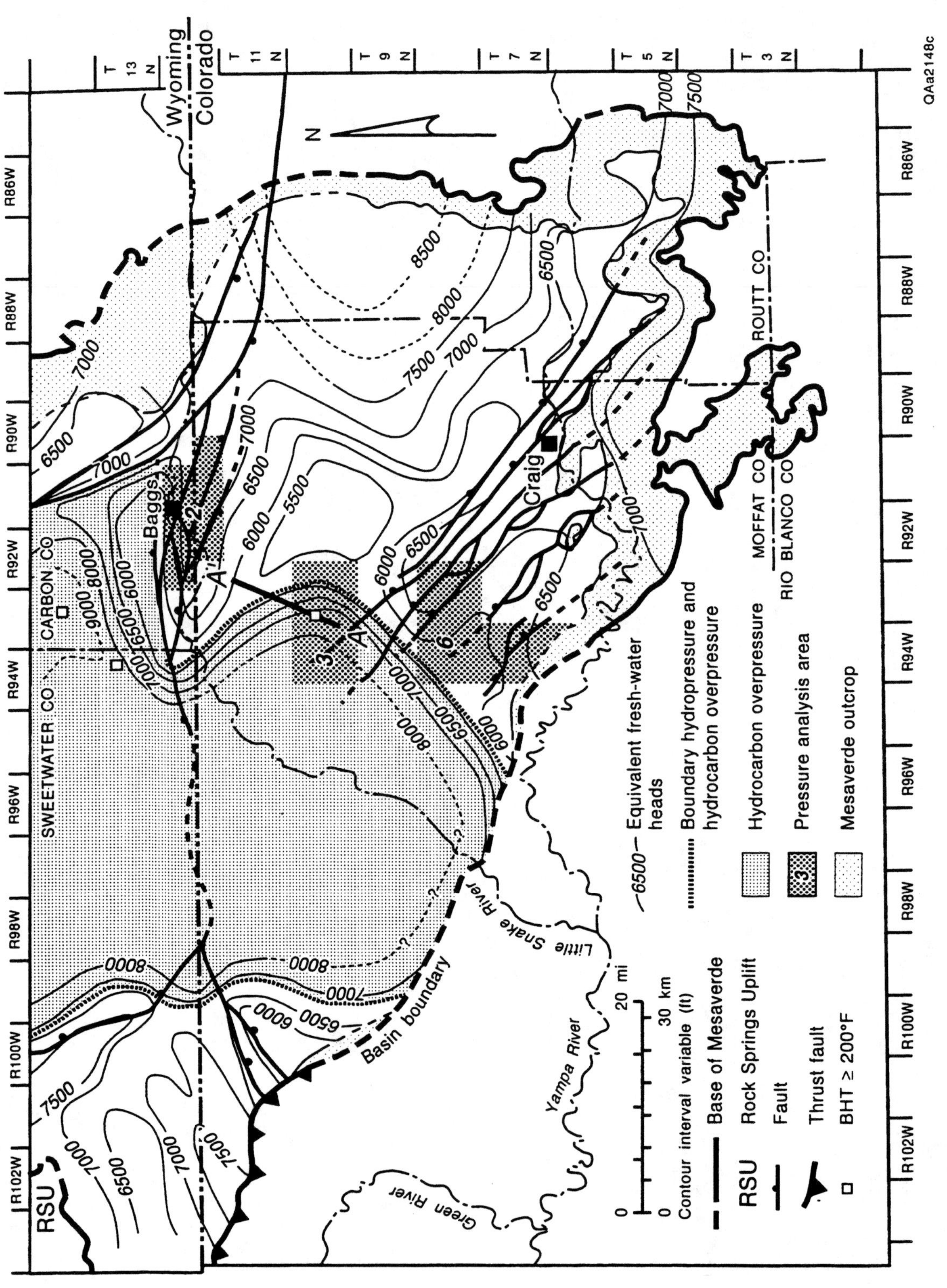

Figure 40. Upper Mesaverde potentiometric-surface map. Ground water flows westward from eastern and southern recharge areas for eventual discharge basinward and to the Yampa River valley. Flow is minor off the southern flank of the Rock Springs Uplift. Simple pressure gradients in pressure-analysis Areas 2, 3, and 6 are 0.48, 0.44, and 0.36 psi/ft (10.9, 10.0, and 8.1 kPa/m), respectively. Cross section A–A′ (fig. 44) shows facies changes associated with the geopressure to hydropressure transition.

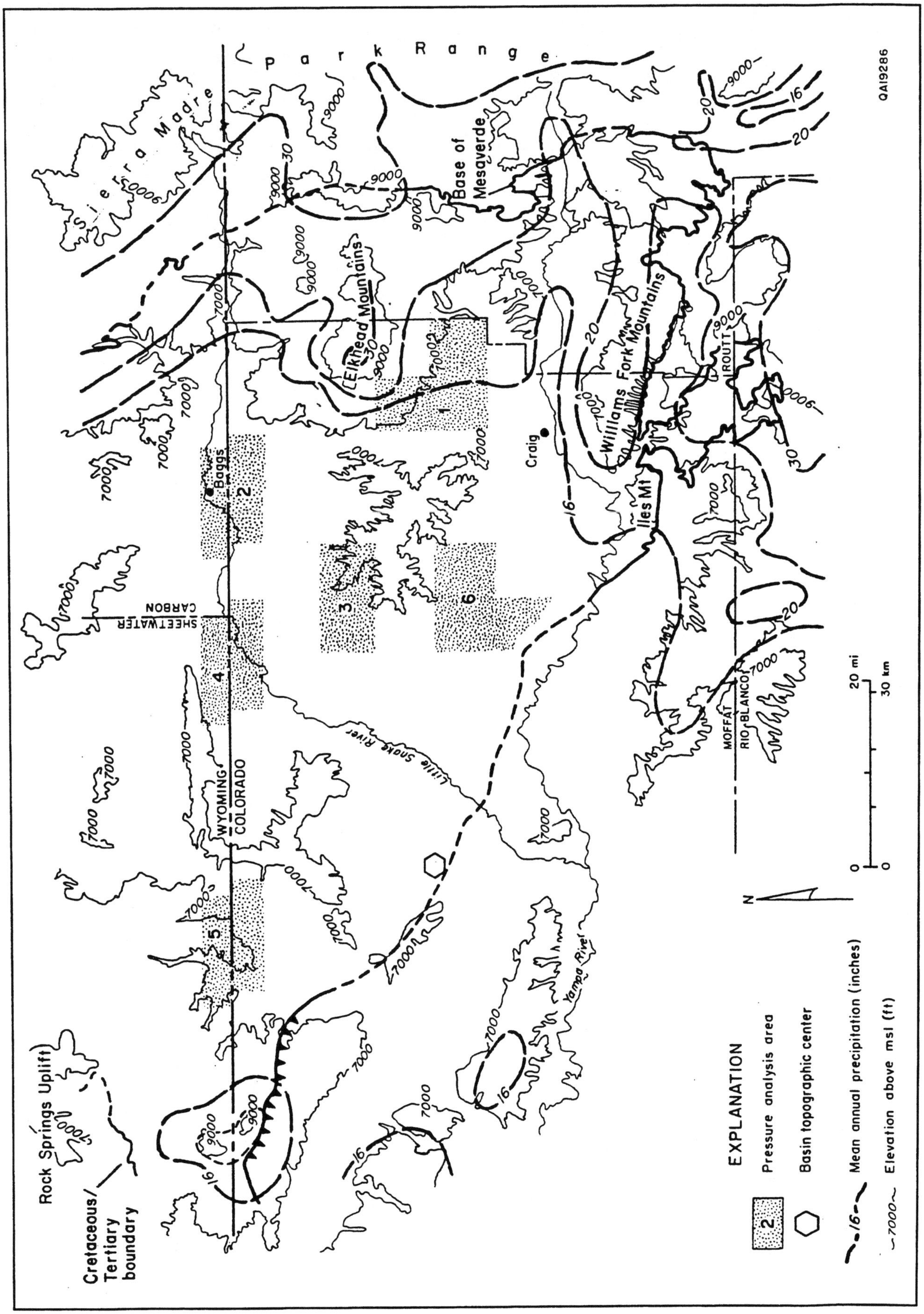

Figure 41. Mean annual precipitation, topography, and major drainage in the Sand Wash Basin. Ground water is recharged over the wet, elevated eastern and southern margins of the basin. Precipitation data from Soil Conservation Service (1983) and Colorado Climate Center (1984).

and burial of the Mesaverde by thrust faults limits recharge from the southwest margin of the basin. Potentiometric highs are also located in the deeper parts of the Sand Wash and Washakie Basins where hydrocarbon overpressure occurs; heads in these areas may exceed 9,500 ft (2,896 m), which is significantly higher than outcrop elevations.

Pressure Regime

Over 300 DST's from 6 study areas (fig. 39) were used to evaluate local variations in pressure regime and to determine potential for vertical flow. These detailed study areas were selected based on DST availability and the geographic distribution of the areas. Only Areas 2, 3, and 6 contained a sufficient number of Mesaverde DST data to fully evaluate simple and vertical pressure gradients within the Mesaverde (fig. 39). Simple pressure gradients (pressure-depth quotients) from these three areas in the Sand Wash Basin indicate that no pressure regime in the hydropressured part of the basin is regionally dominant. However, most gradients indicate slight underpressure to normal pressure. Simple pressure gradients in pressure analysis Areas 2, 3, and 6 are 0.48, 0.44, and 0.36 psi/ft (10.9, 10.0, and 8.1 kPa/m), respectively (fig. 42). Overpressure in Area 2 is artesian in origin and reflects proximity to the recharge area, basinward confinement, aquifer offset by faults along the Cherokee Arch, and high permeability; flowing artesian wells at Dixon field attest to artesian conditions in this area. Overpressure extends approximately 15 mi (~24 km) west of the outcrop to the middle of T12N, R92W. Underpressure in Area 6 may reflect poor connection with the outcrop recharge area or the draining effect of higher permeability downflow such that discharge exceeds recharge, keeping the area underpressured. Although head contours in Area 6 are subparallel to the outcrop, indicating potential for some recharge from the margin (fig. 40), rainfall in this area is significantly lower than the eastern margin of the basin (fig. 41). Furthermore, the area's northwest structural grain may inhibit northeast movement of ground water. A potentiometric valley on the southwest side of the Cedar Mountain Fault system (fig. 40) may reflect sealed or disconnected faults that inhibit the northwest movement of ground water from the southern outcrop and/or the presence of muddy backbarrier/floodplain deposits that occur behind barrier/strandplain deposits northeastward (Hamilton, personal communication, 1993). Moreover, discharge to the Yampa River valley may limit underflow available for recharge basinward to the confined aquifer.

The vertical pressure gradient, which is the slope of the pressure-elevation plot, is used to indicate vertical flow direction. Vertical flow in the Mesaverde is potentially upward in the northern part of the basin and downward in the southern part. The vertical pressure gradient in Area 2 is 0.88 psi/ft (19.9 kPa/m) (fig. 42), well in excess of the hydrostatic pressure gradient (0.43 psi/ft [9.7 kPa/m]), indicating very strong potential for upward flow and poor vertical connectivity (good confinement), which is consistent with overpressured conditions. Vertical gradients in Areas 3 and 6 (0.47 and 0.39 psi/ft [10.6 and 8.8 kPa/m], respectively) indicate weak upward and moderate downward potential for vertical flow, respectively. Downward flow in Area 6 may reflect flow down steep structural dip (Tyler and Tremain, this vol., fig. 5).

Hydrocarbon overpressuring is postulated from head data and bottom-hole temperatures (BHT). Fresh-water equivalent heads exceed 9,500 ft (>2,896 m), which is considerably higher than that of the Mesaverde outcrop on the east, indicating that these anomalously high heads are probably not due to artesian conditions. Simple pressure gradients for the Washakie Basin range from 0.50 psi/ft to more than 0.85 psi/ft (11.3 to >19.2 kPa/m) (McPeek, 1981). Overpressure in the deep Washakie Basin is probably hydrocarbon related where gas is the pressuring fluid rather than water. Overpressure is predicted on low permeability (<0.1 md) and active generation of gas (Law and Dickinson, 1985; Law and others, 1986) and is characterized by pressure gradients greater than 0.70 psi/ft (>15.8 kPa/m). Thus, geopressure and hydropressure are present in the same basin.

Artesian and hydrocarbon overpressures are separated by faults along the eastern margin of the Sand Wash and Washakie Basins (fig. 43). The Savery fault system, which extends northwestward from Savery, Wyoming, and the east–west-trending Cherokee Arch fault system, separate shallow artesian overpressure on the east from hydrocarbon overpressure in the deep Washakie Basin on the west. On the upthrown side of the northwest-trending fault system, overpressured to normally pressured conditions are present where ground water moves basinward from the eastern margin of the basin. In areas where the faults act as barriers to lateral flow, artesian overpressure is generated. On the downthrown side, where the Mesaverde Group is faulted against the Mancos Shale, BHT's approach or exceed 200°F (≥93°C), whereas temperatures on the upthrown side are less than 100°F (<38°C). The southern boundary of hydrocarbon overpressure corresponds to east–west faults along the Cherokee Arch but may be due to diagenetic processes as well. Permeability decreases from approximately 0.6 md in T14N to less than 0.05 md closer to the fault system (fig. 43), suggesting that compactional fluids moving up and out of the basin center were trapped by the fault systems. Therefore, the zone adjacent to the faults may be characterized by intense

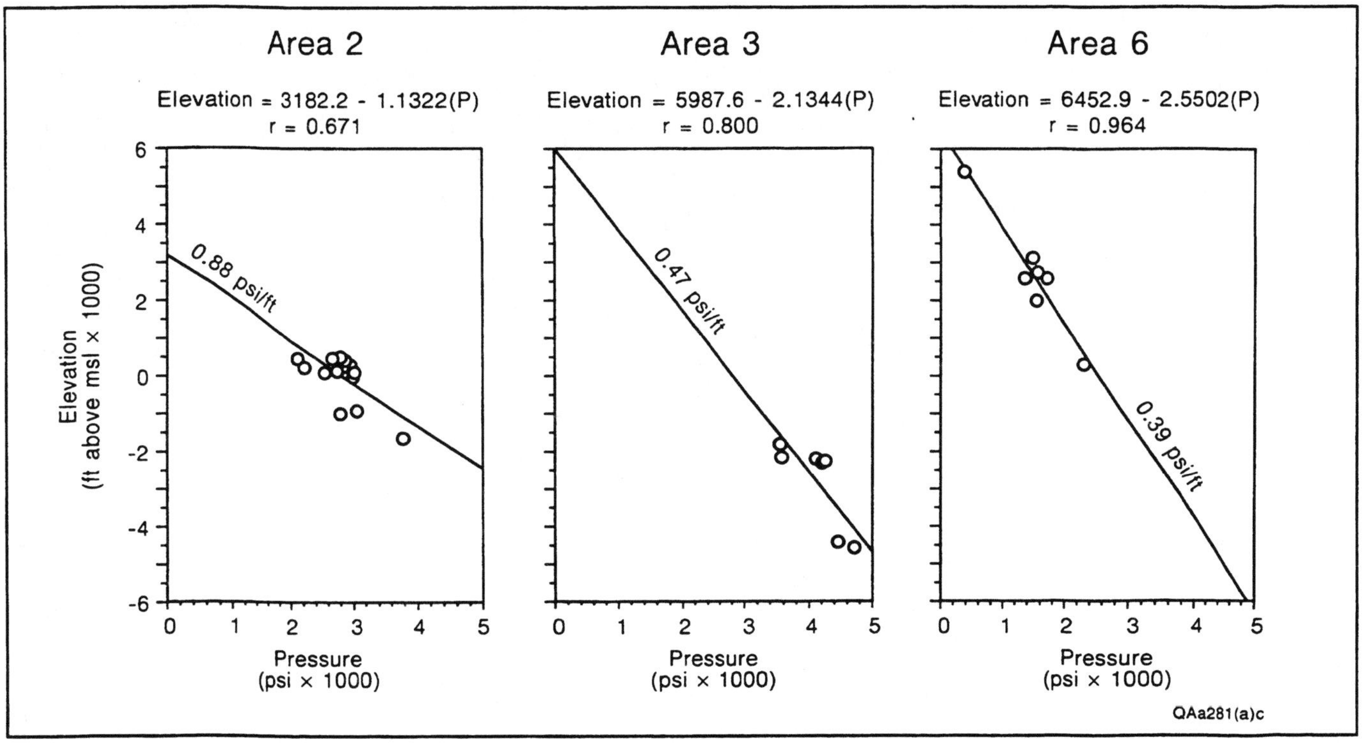

Figure 42. Mesaverde pressure-elevation plots for pressure-analysis, Areas 2, 3, and 6. The large vertical-pressure gradient in Area 2 (~0.88 psi/ft [~19.9 kPa/m]) indicates a strong potential for upward flow and poor vertical connectivity (good confinement), whereas Areas 3 and 6 indicate weak upward and moderate downward-flow potentials, respectively. Downward flow in Area 6 may reflect flow down steep structural dip (Tyler and Tremain, this vol., fig. 5).

diagenesis at the interface between compactional and meteoric waters.

The transition from geopressure to hydropressure in T10–11N, R93–94W and to the south is influenced by facies changes and, probably, diagenetic processes. The thickness of the Almond Formation decreases from more than 300 ft (>91 m) to 0 ft over approximately 12 mi (~9 km) (fig. 44). Almond shoreline sandstones are well developed in T10N, R94W but become thinner and more shaly eastward; net sandstone decreases from more than 220 ft (>67 m) to less than 80 ft (<24 m) over more than 5 mi (>8 km) (fig. 44). A northwest-trending fluvial channel system in Unit 4 corresponds to the transition from geopressure to hydropressure (Hamilton, this vol., fig. 23). The thick Unit 4 fluvial channel sandstones in T10N, R93W grade eastward and northeastward into shaly, floodplain/lacustrine facies (fig. 44), suggesting that this fluvial channel belt may represent the eastward transition from hydrocarbon overpressure to hydropressure in Unit 4. These facies changes may also represent a mixing zone for compactional and meteoric waters. Formation waters moving up and out of the basin center and water moving basinward from the basin recharge areas can mix in this transition zone (fig. 45). Reduction of permeability through cementation would direct fluid vertically and also result in a relatively abrupt transitions between pressure regimes.

Hydrochemistry

Chlorinity and TDS contents are lowest along the eastern and southern Mesaverde outcrop belt, increase westward along the Cherokee Arch, and northwestward from the Craig area (figs. 46 and 47). Chlorinity content is lowest along the eastern and southern Mesaverde outcrop and increases westward along the Cherokee Arch and northwestward from the Craig area (fig. 46). Chlorinities ranged from less than 50 mg/L near the outcrop to more than 61,000 mg/L along the Uinta thrust belt. At Dixon field (T12N, R90W), chlorinities are less than 250 mg/L but increase abruptly to greater than 5,000 mg/L west of Baggs within individual fault blocks along the Cherokee Arch. Northwest of Craig chlorinities are less than 500 mg/L.

The TDS ranges from less than 1,000 mg/L in the eastern outcrop to more than 104,000 mg/L along the Uinta thrust belt (fig. 47). Westward along the Cherokee Arch, TDS increased from less than 5,000 mg/L in the Baggs area to more than 11,000 mg/L in T12N, R92W reflecting reservoir segmentation by faults along the Cherokee Arch. TDS on the upthrown side on the Savery Fault system is less than 3,000 mg/L, whereas TDS on the downthrown side of the fault system is estimated to be 29,000 mg/L (Collentine and others, 1981).

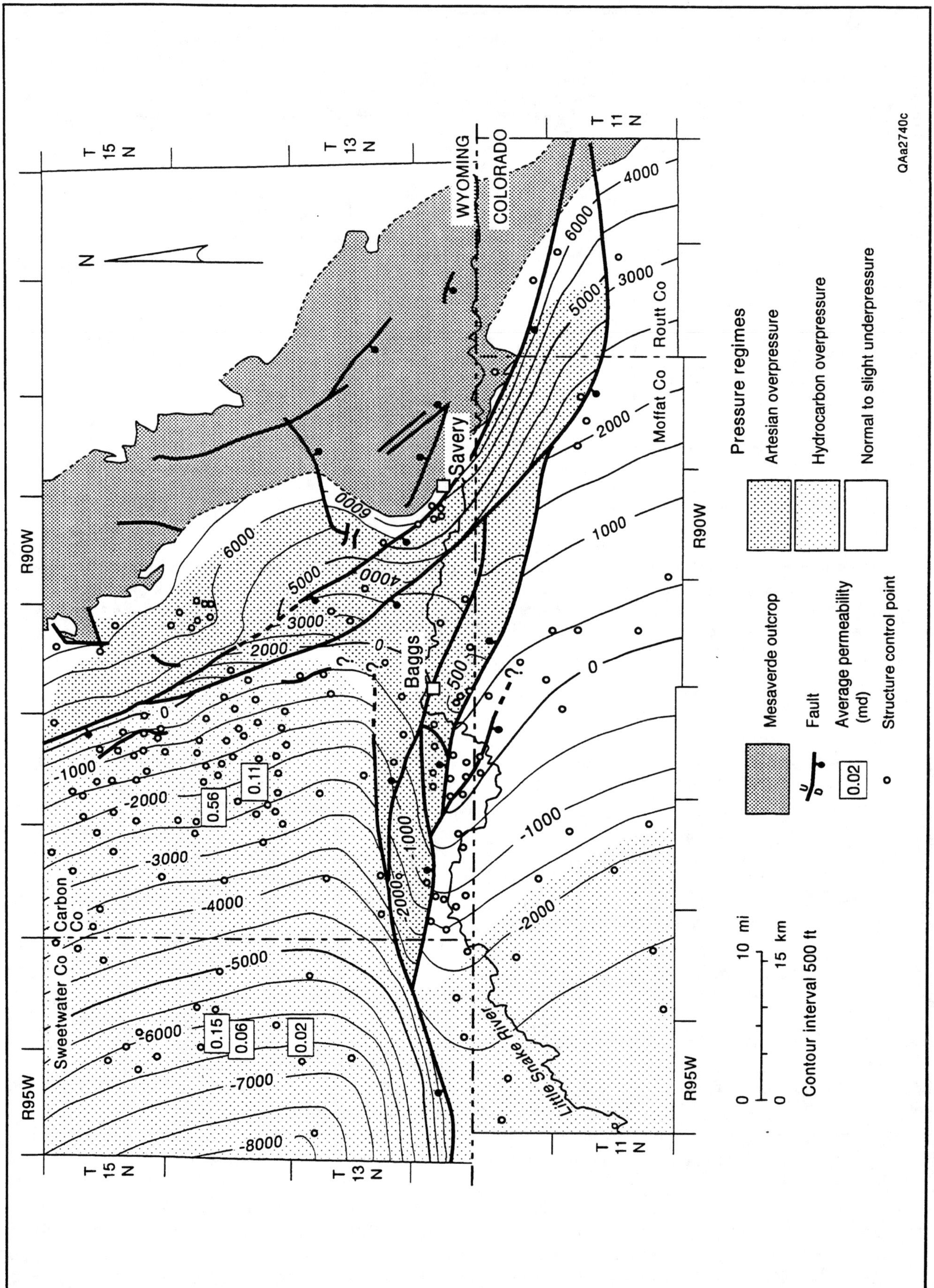

Figure 43. Structure-contour map of the top of the Williams Fork Formation in the Sand Wash and Washakie Basins. Here, the distribution of hydrocarbon and artesian overpressure is controlled primarily by faulting. The average permeability in the producing interval decreases toward faults along the Cherokee Arch.

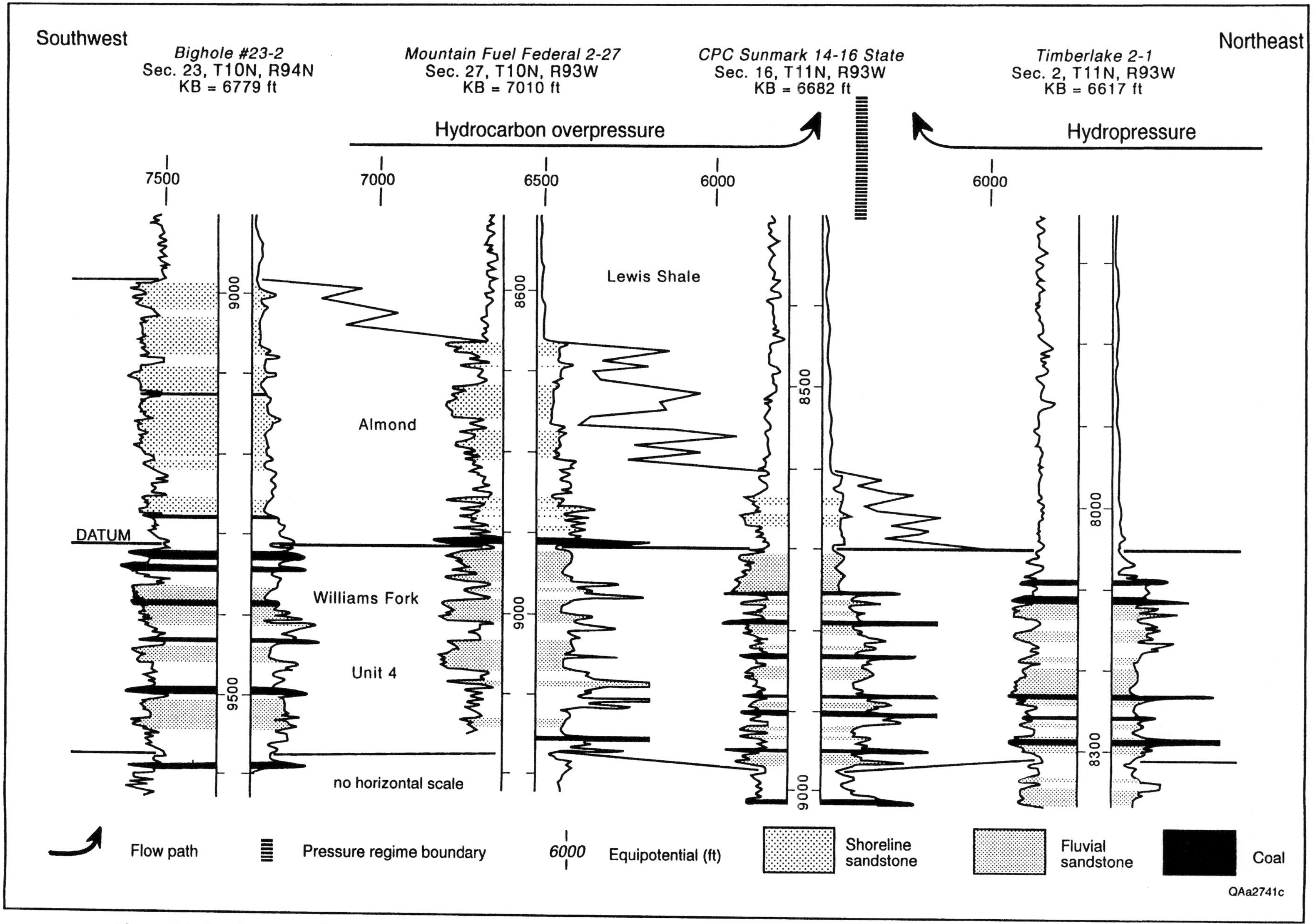

Figure 44. Southwest-northeast cross section across the geopressured-hydropressured transition zone. Pinch-out of Almond shoreline and fluvial sandstones in Unit 4 of the Williams Fork Formation correspond to the transition from geopressure to hydropressure. Location of section is shown in figure 40.

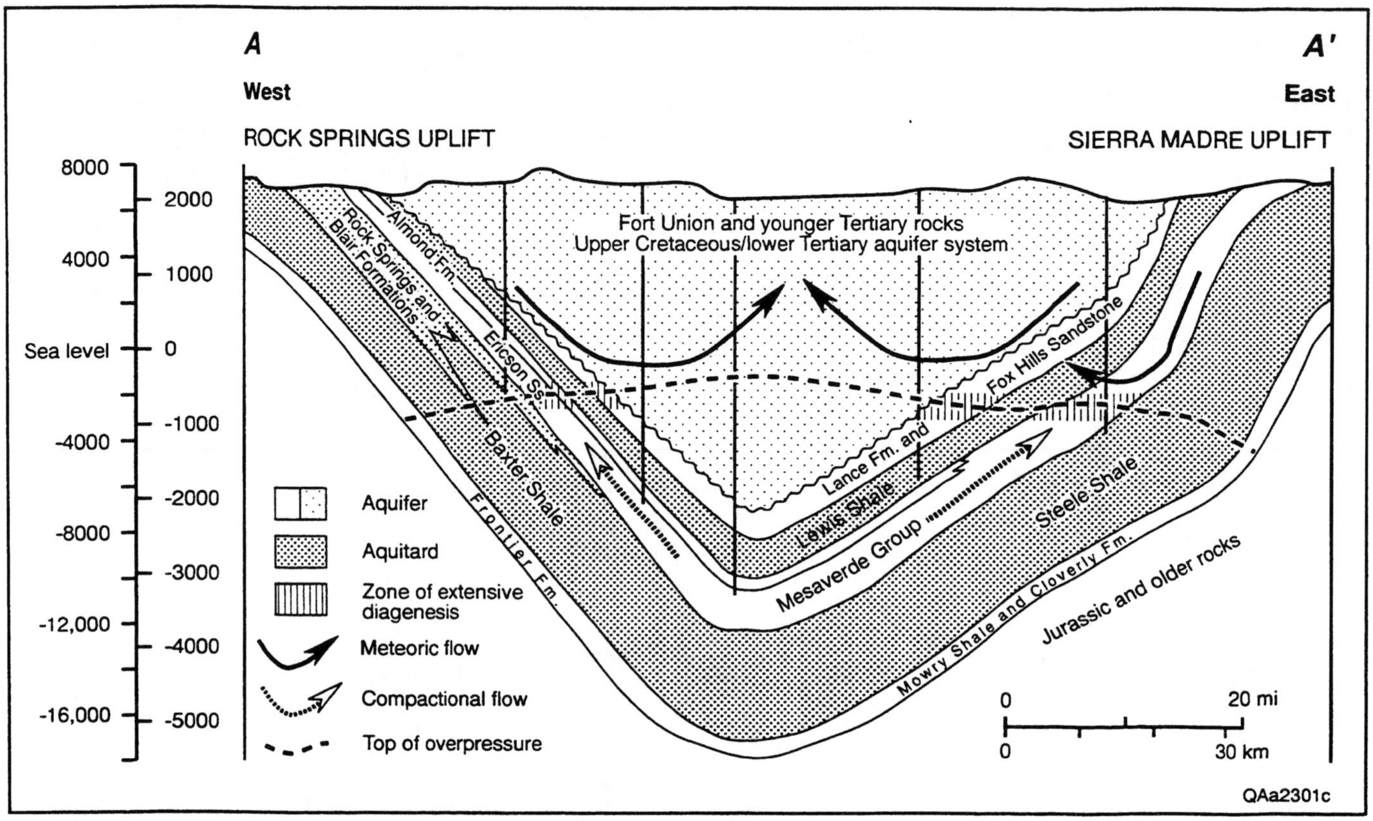

Figure 45. West-east cross section through Washakie Basin (modified from Law and others, [1989]). Recharge is over Sierra Madre and Rock Springs Uplifts. Ground water flows basinward, turning upward upon convergence from the basin margins, aquifer pinch-out, encountering the top of geopressure, or a combination of these. The top of regional overpressure is a no-flow boundary. Line of section is shown in figure 1 of Tyler and Tremain (this vol.).

The TDS is less than 2,000 mg/L along the southern outcrop, generally less than 4,000 mg/L northwest of Craig (T7-8N, R93W), and highest south of the Rock Springs Uplift, where TDS is typically greater than 25,000 mg/L. At Craig Dome field, TDS ranges from 700 to 1,100 mg/L and waters are Na-HCO_3 type.

Regional Flow

Regional ground-water circulation reflects permeability, present-day structural configuration (attitude of aquifers and aquitards), topography, climate (precipitation and infiltration), structural grain, and depositional fabric. Hydrocarbon overpressure and hydropressure are present in the basin. However, these two distinct pressure regimes are always separated by some type of permeability barrier, which may be related to fault zones, areas of intense diagenesis, facies changes, or a combination of these factors. Because of low permeability and high pressure, the top of hydrocarbon overpressure acts as a no-flow boundary. Ground-water flow turns upward upon convergence from the basin margins, aquifer pinch-out, or both, and upon encountering the top of geopressure (fig. 45). Therefore, the basinward movement of ground water is limited by the upward flow of compactional fluids and hydrocarbon overpressure in the deeper parts of the basin. Ground water flows mainly from the east, whereas flow from the west is restricted because of a more arid climate. In the eastern Sand Wash Basin, ground water flows westward down hydraulic gradient, perpendicular to the head contours, in response to the topographic gradient and structural dip (fig. 40; Tyler and Tremain, this vol., fig. 5). In the southeastern part of the basin, ground water flows from the eastern and southern recharge areas toward the Yampa River valley east of Craig for eventual discharge (fig. 40). The chlorinity and TDS gradients show that ground water flows westward and northwestward from the eastern and southern margins of the basin, respectively (figs. 46 and 47).

In Unit 4 thick shoreline sandstones having net sand thicknesses exceeding 70 ft (>21 m) (Hamilton, this vol., fig. 23) occur along the eastern margin of the basin, where the highest annual precipitation occurs. The potentiometric surface steepens basinward (fig. 40), where the net sandstone decreases as the shoreline sandstones grade into lacustrine and fluvial facies. The

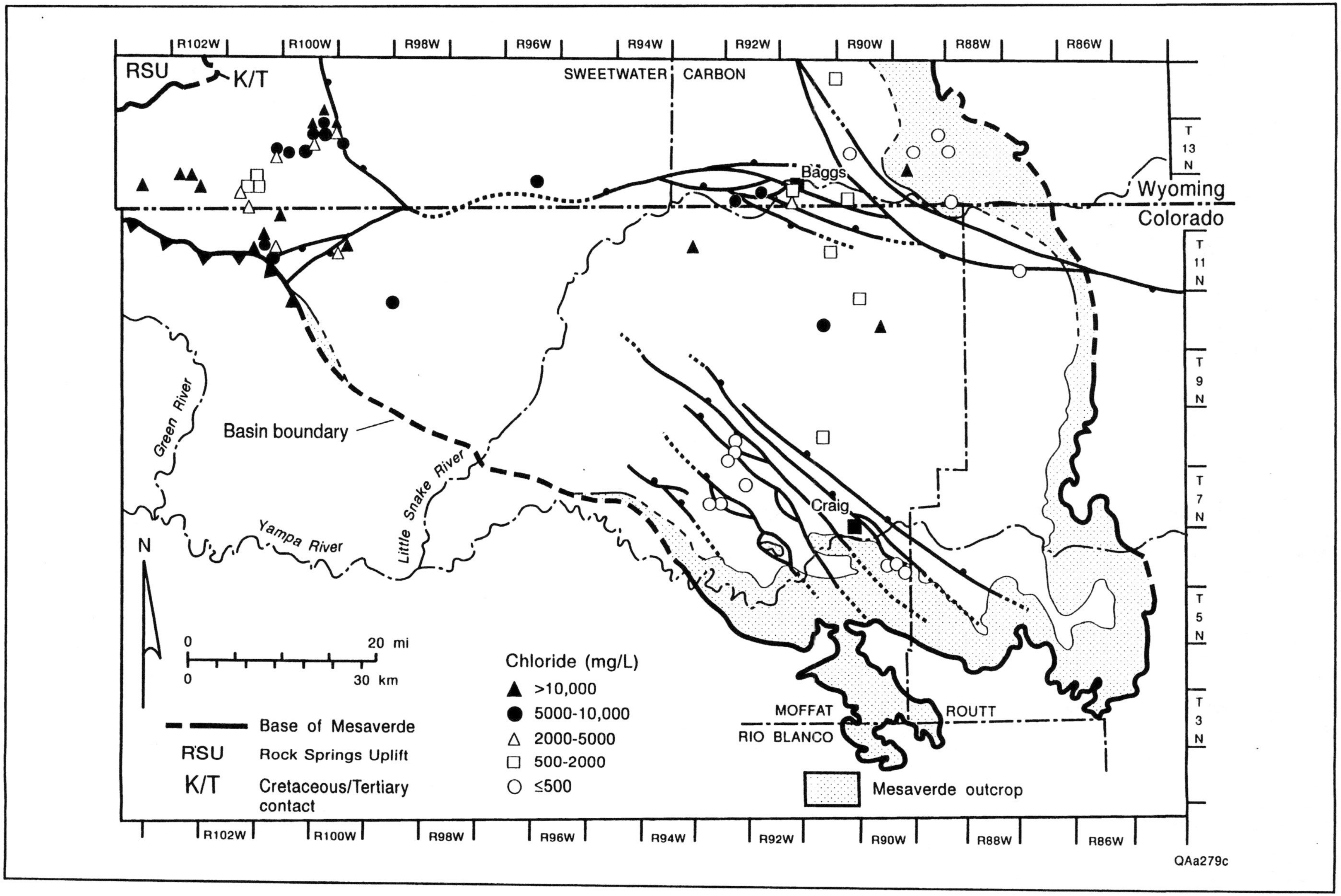

Figure 46. Mesaverde chlorinity map. Chlorinity of formation waters ranges from less than 100 mg/L east of Baggs to generally more than 10,000 mg/L in the western part of the basin. The chlorinity gradient shows that ground water flows westward and northwestward from the eastern and southern margins of the basin, respectively.

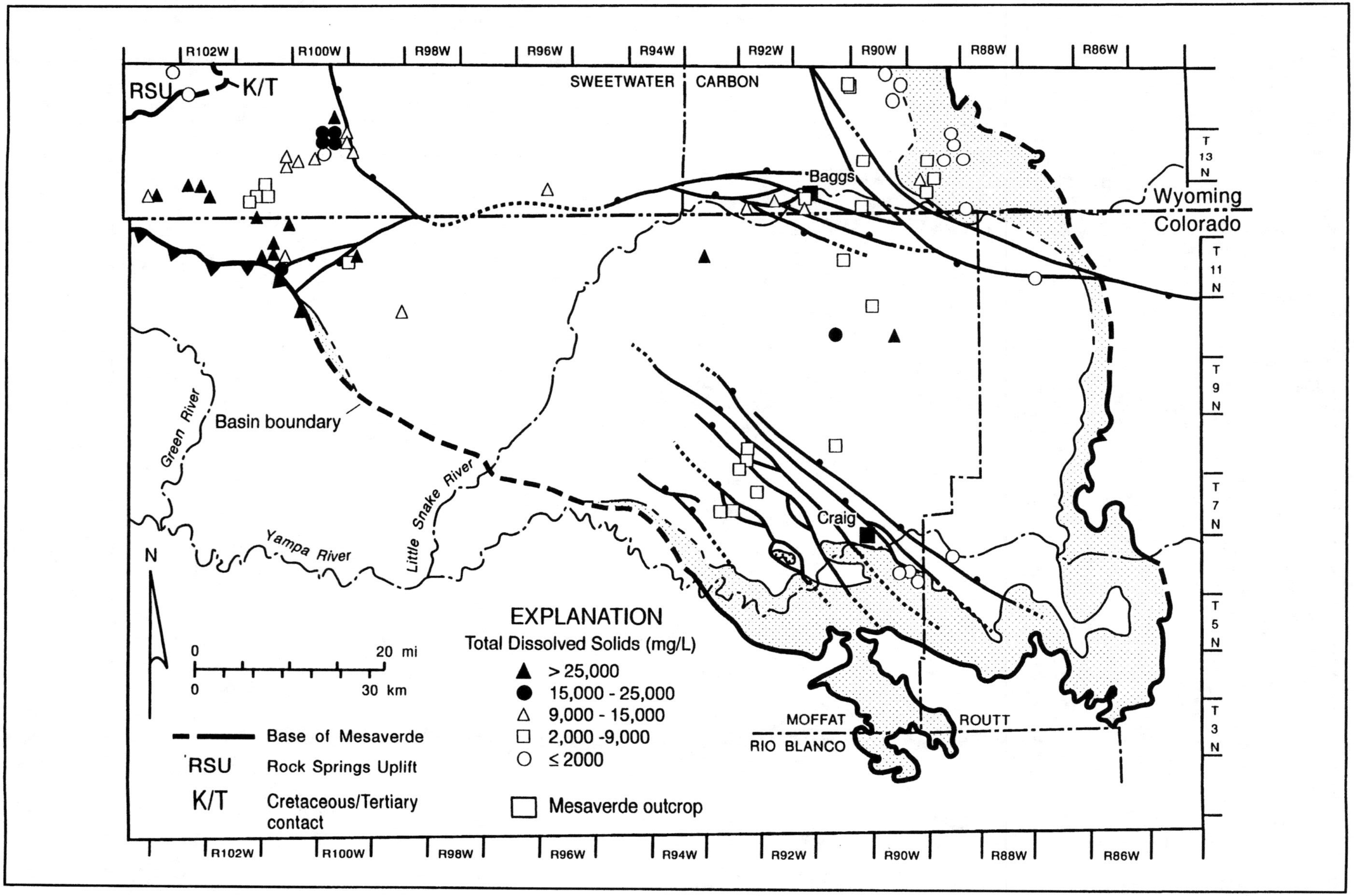

Figure 47. Mesaverde TDS map. The TDS are generally less than 2,000 mg/L along the eastern and southern outcrops and increase to more than 104,000 mg/L near the Uinta Thrust Belt south of the Rock Springs Uplift. The TDS increases from approximately 5,000 mg/L in the Baggs area to more than 11,000 mg/L in T12N, R92W. The TDS gradient indicates that ground water flows basinward from Mesaverde outcrops.

QAa873c

potentiometric nose in T8N, R91W corresponds to a southwestward-trending fluvial-channel belt (Hamilton, this vol., fig. 23), indicating that basinward movement of water is influenced by facies distribution in addition to topographic gradient and structural dip. Potentiometric lows in the eastern Sand Wash Basin generally coincide with areas of low net sand thickness. Although coal beds may be the most important aquifers in the Mesaverde Group because of their high permeability and lateral continuity, coal beds in Unit 4 are thin and discontinuous near the eastern margin of the basin, and depositional dip (northwest to southeast) is perpendicular to basinward water movement. Therefore, the sandstones in Unit 4 are probably more important aquifers than the coals along the eastern margin of the basin. However, northeast-trending shoreline sands and coal beds in the lower Williams Fork (Units 1 and 2) are generally oriented parallel to subparallel to hydraulic gradient. Thus, ground-water flow in these units probably has a strong southwest component. Coal beds in Units 1 and 2 formed behind the shoreline sandstones and are continuous over large areas of the basin (Hamilton, this vol.). Therefore, the coal beds in these units may be the most important Mesaverde aquifers because of their relatively high permeability (50 to 1,462 md) and lateral continuity.

Fracture flow to the west and northwest along the Cherokee Arch and Cedar Mountain fault systems, respectively, is indicated by potentiometric ridges along the fault zones and hydrochemistry. Equivalent freshwater heads on the Cherokee Arch exceed 6,500 ft (1,982 m) and extend westward to R93W approximately 20 mi (~32 km) from the outcrop (fig. 40). Although fracture flow may be promoted westward along the complexly faulted Cherokee Arch (Tyler and Tremain, this vol., fig. 5), faulting may also serve to compartmentalize the aquifer and actually impede westward flow and limit the extent of artesian overpressure. Chlorinities are generally less than 1,000 mg/L near Baggs, Wyoming, but increase abruptly to more than 5,000 mg/L within individual fault blocks west of Baggs (fig. 46). The north-trending Savery Fault system also inhibits the westward flow of ground water; note the potentiometric low on the downthrown side (fig. 40). Moreover, TDS on the upthrown side of the fault is less than 3,000 mg/L, whereas on the downthrown side of the fault in T15N, R92W it is 29,000 mg/L (Collentine and others, 1981). Ground-water flow near the fault may be directed vertically if the fault is a flow barrier, or laterally into the Fort Union, Lance, or Fox Hills Formations if there are permeable holes along the fault plane (Smith and others, 1990). Because of low permeability in coal beds at depths greater than 6,000 to 7,000 ft (>1,829 to 2,134 m) on the downthrown side of the Savery Fault system, this fault system represents the westward limit for coalbed methane exploration in the northeastern Sand Wash and eastern Washakie Basins.

Heads along the Cedar Mountain fault system also exceed 6,500 ft (1,982 m) and a potentiometric ridge extends more than 20 mi (>32 km) from the outcrop (fig. 40). Although there is more than 5,000 ft (>1,524 m) of displacement across the fault system (Tyler and Tremain, this vol., fig. 5), the potentiometric surface does not change significantly across the fault system, indicating that pressure head increases northeastward across the fault system. This increase in pressure head is probably due to fracture flow. The orientation and deposition of Almond shoreline sandstones were influenced by the Cedar Mountain Fault system (Hamilton, this vol.). Ground-water flow from the southern recharge area in the Williams Fork Mountains may be partially controlled by the northwestward orientation of both the Almond sandstones and the fault system. However, coalbed permeability is probably higher than sandstone permeability, suggesting that flow would be preferentially focused in the coal beds. The northwest-trending Cedar Mountain fault system also coincides with the highest net coal thickness trends (net coal greater than 40 ft [>12 m]; Hamilton, this vol., fig. 24) in Unit 4, suggesting that northwest flow may be further promoted by northwest-trending face cleat strikes (Tyler and Tremain, this vol., fig. 4) and the dominant northwest structural grain. Northwest flow through the coal beds would occur within individual fault blocks. Basinward flow in Unit 4 may also be partly controlled by the orientation of fluvial channel belts. The potentiometric surface on the southwest side of the Cedar Mountain Fault system decreases and then increases again near the outcrop (fig. 40). This decrease in the potentiometric surface corresponds with a decrease in net sand thickness in Unit 4, whereas the increase in the potentiometric surface near the outcrop corresponds with an increase in net sand thickness. Net sandstone for the Almond Formation also decreases on the southwest side of the fault system (Hamilton, this vol., fig. 26) suggesting that the potentiometric valley may reflect the transition from barrier to back-barrier depositional setting. Furthermore, while northwest-trending face-cleat strikes may promote northwest flow in this area, other cleat orientations to the west may impede northeast flow from the outcrop (Tyler and Tremain, this vol., fig. 4) and further contribute to underpressure in pressure analysis Area 6 (fig. 40).

Limited recharge over the Rock Springs Uplift is indicated by the high-chlorinity (>5,000 mg/L) formation waters (fig. 46). However, chlorinities generally less than 3,000 mg/L and a potentiometric surface higher than 7,000 ft (2,134 m), extending northeast in T12N, R101W (Wyoming), suggest that recharge does occur from the southwest margin of the basin. Adjacent highlands have elevations greater than 9,500 ft (>2,896 m) and annual precipitation exceeds 16 inches (41 cm) per year (fig. 41). Ground water flows northwest from the potentiometric ridge and southward from the Rock Springs Uplift toward a potentiometric low in T13N, R102W (fig. 40). However, ground water from the southeastern flank of the anticline and southeastern flank of the Rock Springs Uplift is directed eastward (fig. 40). The eastward movement of the water is inhibited by major northwest- and southwest-trending faults and hydrocarbon overpressure.

Conclusions

1. The Mesaverde Group is a thick, regionally interconnected aquifer system of high transmissivity confined by the marine Lewis and Mancos Shales. Coal beds may be the most permeable aquifers, having permeabilities of tens to thousands of millidarcys.

2. Recharge is at outcrop along the wet, elevated eastern and southeastern margins of the basin in the foothills of the Sierra Madre and Park Uplifts and the Williams Fork Mountains, where annual precipitation exceeds 20 inches/yr (50 cm/yr). Lesser amounts of recharge are received over the south end of the Rock Springs Uplift and the foothills of the Uinta Mountains.

3. Ground water flows westward and northwestward, respectively, from an eastern and southern recharge area for eventual discharge, to the east-central part of the basin, and to the Yampa River valley east of Craig, Colorado. Flow direction is influenced by depositional fabric and northwest-trending structural grain. The chlorinity and TDS gradients reflect the regional flow direction.

4. A geopressured and a hydropressured system are recognized in the deep and shallow basins, respectively. Hydrocarbon overpressure and hydropressure are hydraulically separated by some type of permeability barrier such as fault zones, areas of intense diagenesis, facies changes, or some combination of these. The transition between geopressure and hydropressure is often abrupt. Because of low permeability, the top of geopressure is a floor for coalbed methane exploration. The hydropressured section is underpressured, normally pressured, and overpressured. Artesian overpressure is present along the eastern part of the Cherokee Arch. No pressure regime in the hydropressured part of the basin is regionally dominant, although vertical pressure gradients indicate slight underpressure to normal pressure.

Paleocene Fort Union Formation

Stratigraphy and Coal Occurrence of the Paleocene Fort Union Formation, Sand Wash Basin

Roger Tyler and R. G. McMurry

Abstract

In the Sand Wash Basin, the Paleocene Fort Union Formation is defined as strata between the massive Upper Cretaceous and lower Tertiary (K/T) sandstone unit and the Eocene Wasatch Formation. Upper Cretaceous and lower Tertiary sedimentation within the Sand Wash Basin is the result of syntectonic sedimentation and Laramide basement thrusting. Characteristic syntectonic sedimentary facies in the Sand Wash Basin include a narrow conglomerate facies adjacent to the thrust, a narrow sandstone/mudstone/coal facies just basinward, a basinal thrustward-thickening mudstone facies, and a wide distal sandstone/mudstone/coal facies. The Paleocene Fort Union Formation is the major coal-bearing unit and coalbed methane target in Tertiary-age rocks of the Sand Wash Basin. The Fort Union Formation is further operationally divided into the lower coal-bearing unit, the gray-green mudstone unit, the basin sandy unit, and the upper shaly unit.

Coal thickness and coal-seam continuity are greatest in the lower coal-bearing unit in the eastern and southeastern Sand Wash Basin. Coals were deposited along predominantly north-flowing fluvial systems, where thick sandstone sequences served as platforms for peat accumulation. Lithofacies and coal-occurrence maps show that maximum coal development corresponds to floodplain deposits above, and on the flanks of, south-north-oriented fluvial systems; individual seams have maximum thicknesses of 20 to 50 ft (6.1 to 15.2 m) and a combined maximum net-coal thickness of approximately 80 ft (~15.2 m). Thinner coal beds (3 to 10 ft thick [0.9 to 3.1 m]) also occur in the western Sand Wash Basin, in both the lower coal-bearing and upper shaly units.

Introduction

One of the problems in regional subsurface correlations of the Paleocene Fort Union Formation in the Sand Wash Basin is the uncertainty of lithostratigraphic boundaries. Different sources of clastic material involving similar or different rock types, mixed environments of deposition (for example, coarse clastics deposited into a floodplain, paludal, or lacustrine environment) and unconformities complicated the stratigraphy of the Fort Union Formation (Masters [1961], Colson [1969], and Beaumont [1979]). To correlate the major coal-bearing horizons in the Fort Union Formation of the Sand Wash Basin, operational lithostratigraphic zones in the Upper Cretaceous and lower Tertiary rocks were defined. These lithostratigraphic zones include the Fox Hills Sandstone, Lance Formation, massive Cretaceous and Tertiary (K/T) sandstone unit, Fort Union Formation, and Wasatch Formation (fig. 48). Similar Upper Cretaceous and lower Tertiary lithostratigraphic zones have been identified by Colson (1969), Beaumont (1979), Honey and Roberts (1989), Hettinger and others (1991), and Hettinger and Kirschbaum (1991). The Fox Hills Sandstone/Lance Formation couplet is depositionally equivalent and homotaxial to the Pictured Cliffs Sandstone/Fruitland Formation couplet, a prolific coalbed methane producer in the San Juan Basin.

In Upper Cretaceous and early Tertiary stratigraphic cross sections through the Sand Wash Basin, the Upper Cretaceous continental Mesaverde Group is overlain by deposits of the offshore marine Lewis Shale, nearshore-marine and marginal-marine Fox Hills Sandstone, and continental fluvial Lance Formation of latest Cretaceous age (figs. 49–52). The overlying fluvial, massive K/T sandstone unit is the host of a regional unconformity that separates Cretaceous from Tertiary-age rocks. The massive K/T sandstone unit intertongues with the underlying Lance Formation and the overlying fluvial Paleocene Fort Union Formation. In the western Sand Wash Basin, the uplift and erosion of parts of the Lewis Shale, Fox Hills Sandstone, and Lance Formation result in an angular unconformity between the Fort Union Formation and the underlying sediments. In the central and eastern Sand Wash Basin, the sediments above and below the regional unconformity appear to be disconformable.

The major coal-bearing units and coalbed methane targets are in the overlying Fort Union Formation, which consists of south-north-oriented fluvial systems that control the occurrence and position of thick sandstone and coal-bearing sequences. The sediment source of the Fort Union Formation fluvial system was predominantly from the Sawatch Range to the south, although

In **Kaiser, W. R., and others, 1993, Geologic and hydrologic controls on coalbed methane: Sand Wash Basin: The University of Texas at Austin, Bureau of Economic Geology, topical report prepared for the Gas Research Institute under contract no. 5091-214-2261 (GRI–92/0420), p. 79-106.**

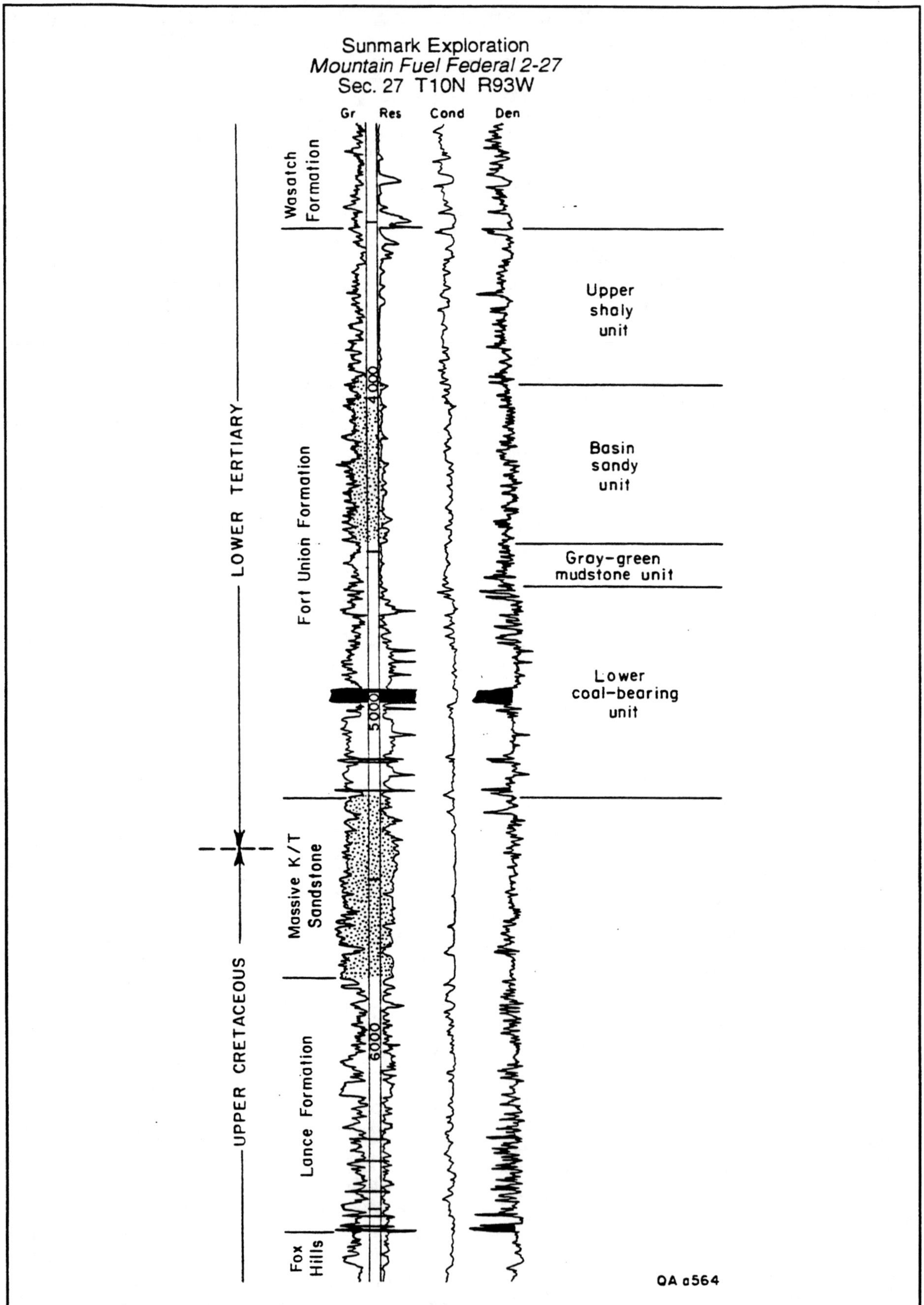

Figure 48. Type log and stratigraphic nomenclature of the Paleocene Fort Union Formation, Sand Wash Basin, and the occurrence of coal. Coal beds identified from density and sonic logs. See figure 52 for location.

several tributaries flowed from the Sierra Madre–Park Uplift to the east and Uinta Uplift to the west. The Fort Union Formation can be further operationally divided into the lower coal-bearing unit, the gray-green mudstone unit, the basin sandy unit, and the upper shaly unit (figs. 48–51). The lower coal-bearing unit is the major coal-bearing and coalbed methane target in the Sand Wash Basin; the upper shaly unit is a minor coalbed methane target.

The fluvial Wasatch Formation, which overlies the upper shaly unit of the Fort Union Formation, records further uplift of the margins of the Sand Wash Basin, as well as continued subsidence within the basin. The contact between the Fort Union and the Wasatch Formations is difficult to recognize in surface and subsurface studies; it generally appears to form an angular unconformity along the western and eastern margins of the Sand Wash Basin but is disconformable in the deep, central parts of the basin.

Lithostratigraphic Zones and Units

Fox Hills Sandstone

The Fox Hills Sandstone was deposited in nearshore-marine and marginal-marine environments during the final regressive phase of the Western Interior Seaway. Nearshore-marine and marginal-marine deposits of the Fox Hills Sandstone intertongue with offshore marine deposits of the underlying Lewis Shale and continental deposits of the overlying Lance Formation (Gill and others, 1970). The upper contact of the Fox Hills Sandstone with the Lance Formation is placed on top of the highest regressive marine sandstone. In the eastern Sand Wash Basin, the progradational Fox Hills Sandstone is about 200 ft (~60.9 m) thick and consists of superimposed upward-coarsening sequences that begin with shale and coarsen upward into thick sandstone bodies, recognized on geophysical logs by their blocky log signatures (figs. 49–51). When traced west these sequences consist of interbedded marine and nonmarine units, which may be partially or completely eroded by the Cretaceous and Tertiary unconformity (figs. 49–51). Thin coal beds from 1 to 2 ft (0.31 to 0.61 m) thick are commonly present in aggradational facies of the Fox Hills to the west. The Fox Hills Sandstone is a very minor coalbed methane target.

Lance Formation

Fluvial deposits of the Lance Formation conformably overlie and intertongue with the Fox Hills Sandstone in the central and eastern Sand Wash Basin (figs. 49–51). The formation is 800 to 1,000 ft (244 to 304.3 m) thick in the southeast and 200 ft (60.9 m), or less, in the northwest. In the western part of the basin, the Lance Formation thins dramatically as a result of erosional truncation by the overlying Cretaceous and Tertiary unconformity (figs. 49–51). The Lance Formation is characterized throughout the basin by multistoried channel-fill sandstone bodies and thin interbedded shale and coal beds. The formation can be subdivided into lower and upper units on the basis of an increased abundance and thickness of the sandstones in the upper unit, recognized by blocky signatures on geophysical logs (figs. 49 and 50). The lower unit thins from about 500 ft (~152 m) in the east to less than 100 ft (<30.6 m) in the west. The basal 150 to 200 ft (458 to 60.9 m) of the lower unit usually contains from one to five lenticular coal beds, 1 to 10 ft (0.31 to 3.05 m) thick (figs. 50 and 51). Locally, these coal beds merge into single seams that are 15 to 20 ft (4.6 to 6 m) thick but are laterally discontinuous. In the eastern Sand Wash Basin, second and third coal packages are sometimes present about 250 and 500 ft (~76 and 152 m) above the base of the formation (fig. 50). These packages contain one or two discontinuous coal beds 1 to 3 ft (0.3 to 0.9 m) thick. The upper unit of the Lance Formation thins from about 600 ft (~183 m) in the east to less than 100 ft (<30.5 m) in the west (figs. 49 and 50). The upper unit consists of laterally discontinuous sandstone sequences ranging from 20 to 100 ft (6.1 to 30.5 m) thick that are separated by shale and mudstone layers 10 to 20 ft (3.05 to 6.1 m) thick. The Lance Formation is a minor coalbed methane target.

Massive Cretaceous and Tertiary (K/T) Sandstone Unit

An interval dominated by thick sandstone sequences overlies and intertongues with the upper zone of the Lance Formation and is overlain by and intertongues with the lower coal-bearing unit of the Fort Union Formation (figs. 49–51). This sequence of rock, referred to as the massive K/T sandstone unit (unnamed Cretaceous and Tertiary sandstones of Hettinger and others [1991] and Ohio Creek of Irwin [1986]), is recognized on geophysical logs by its blocky log signature, thickness in hundreds of feet, and stratigraphic position below the coal-bearing Fort Union Formation (figs. 49–51). The blocky signatures are correlatable throughout the basin and north into the Washakie Basin (Hettinger and others, 1991).

The massive K/T sandstone unit is interpreted as a large, north-flowing, low-sinuosity bedload fluvial system. The prominent tributary pattern displayed on the net-sandstone-thicknesses map (fig. 53) defines

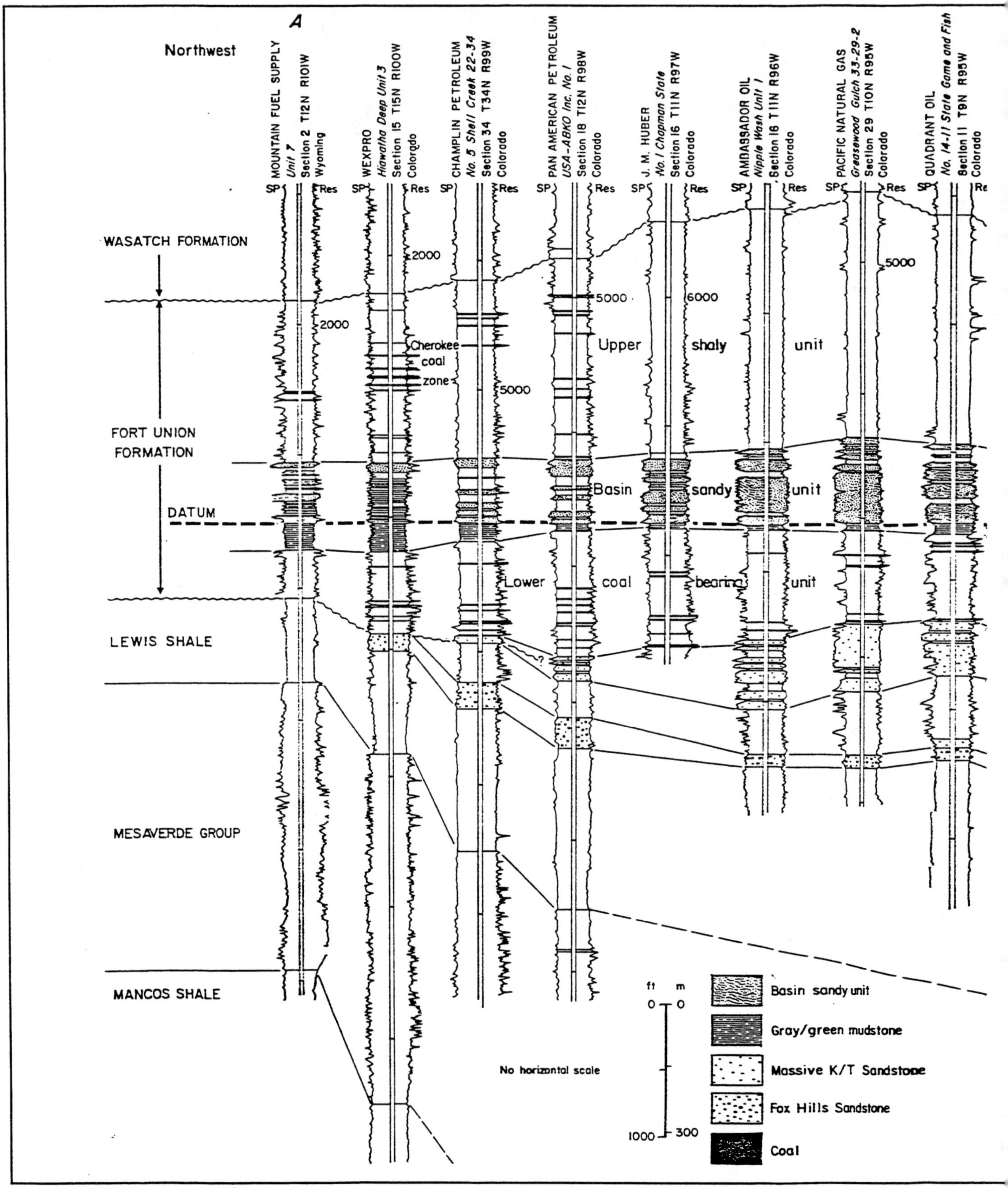

Figure 49. Northwest-southeast stratigraphic cross section A–A′ of the Paleocene Fort Union Formation, Sand Wash Basin, illustrating operationally defined stratigraphic units and coal occurrence. Thick coal beds occur in the eastern Sand Wash Basin in the lower coal-

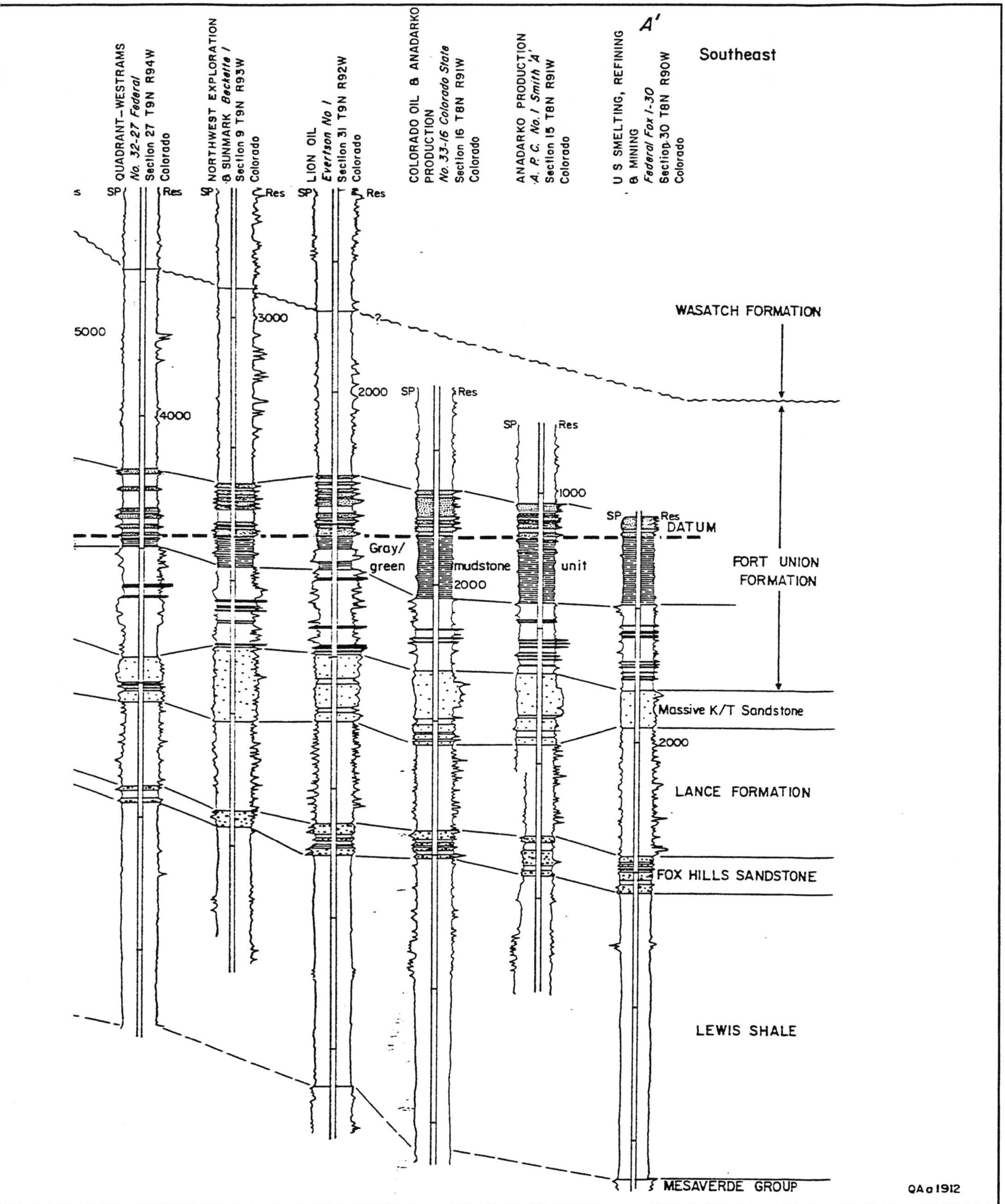

bearing unit and in the northwestern Sand Wash Basin in the upper shaly unit. Coal beds are best developed above thick fluvial sandstones.

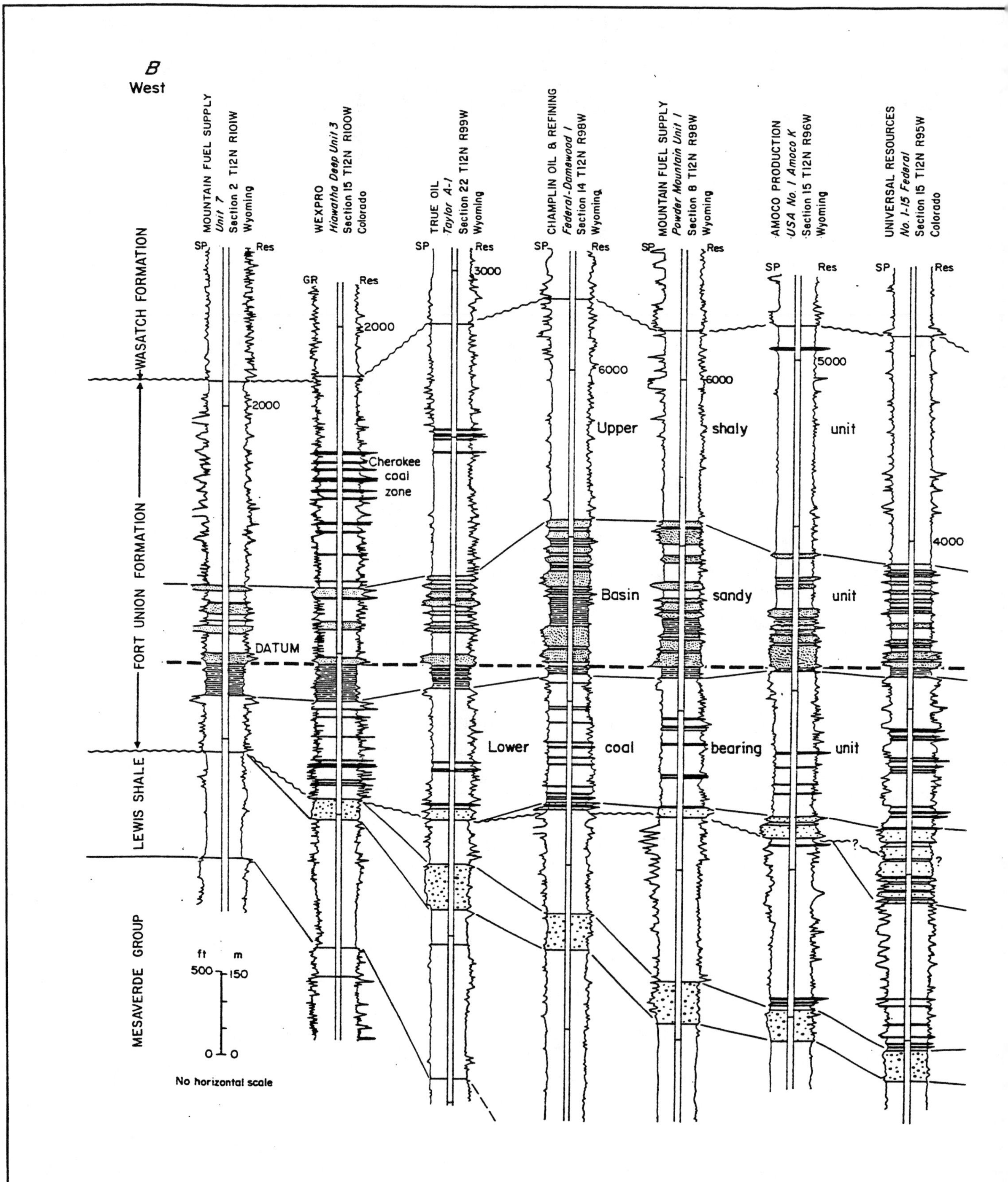

Figure 50. West-east stratigraphic cross section B–B′ through T12N of the Paleocene Fort Union Formation, Sand Wash Basin, illustrating operationally defined stratigraphic units

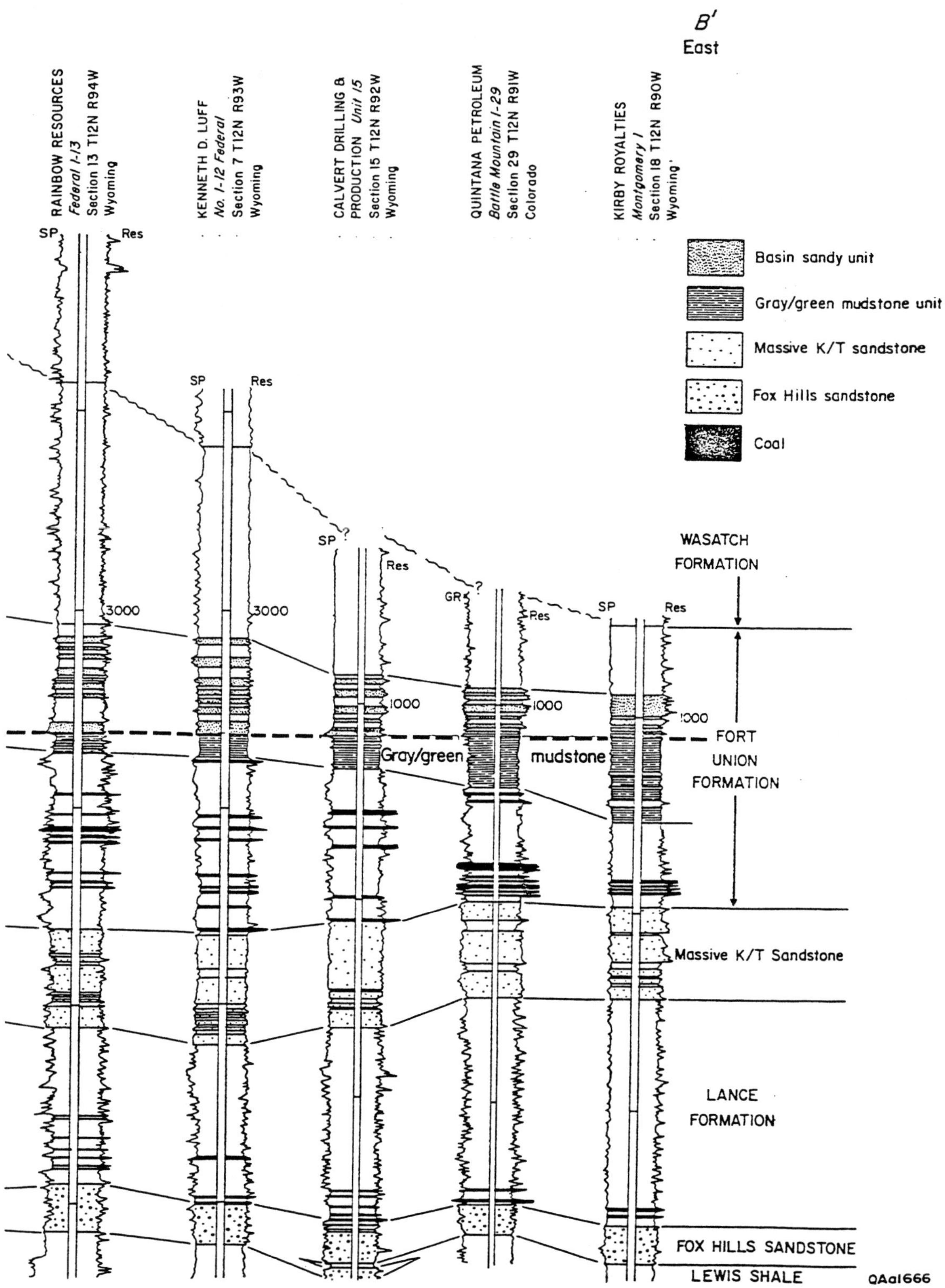

and coal occurrence of the lower coal-bearing unit. Thickest and more continuous coal beds occur in the eastern Sand Wash Basin above thickest sandstone development.

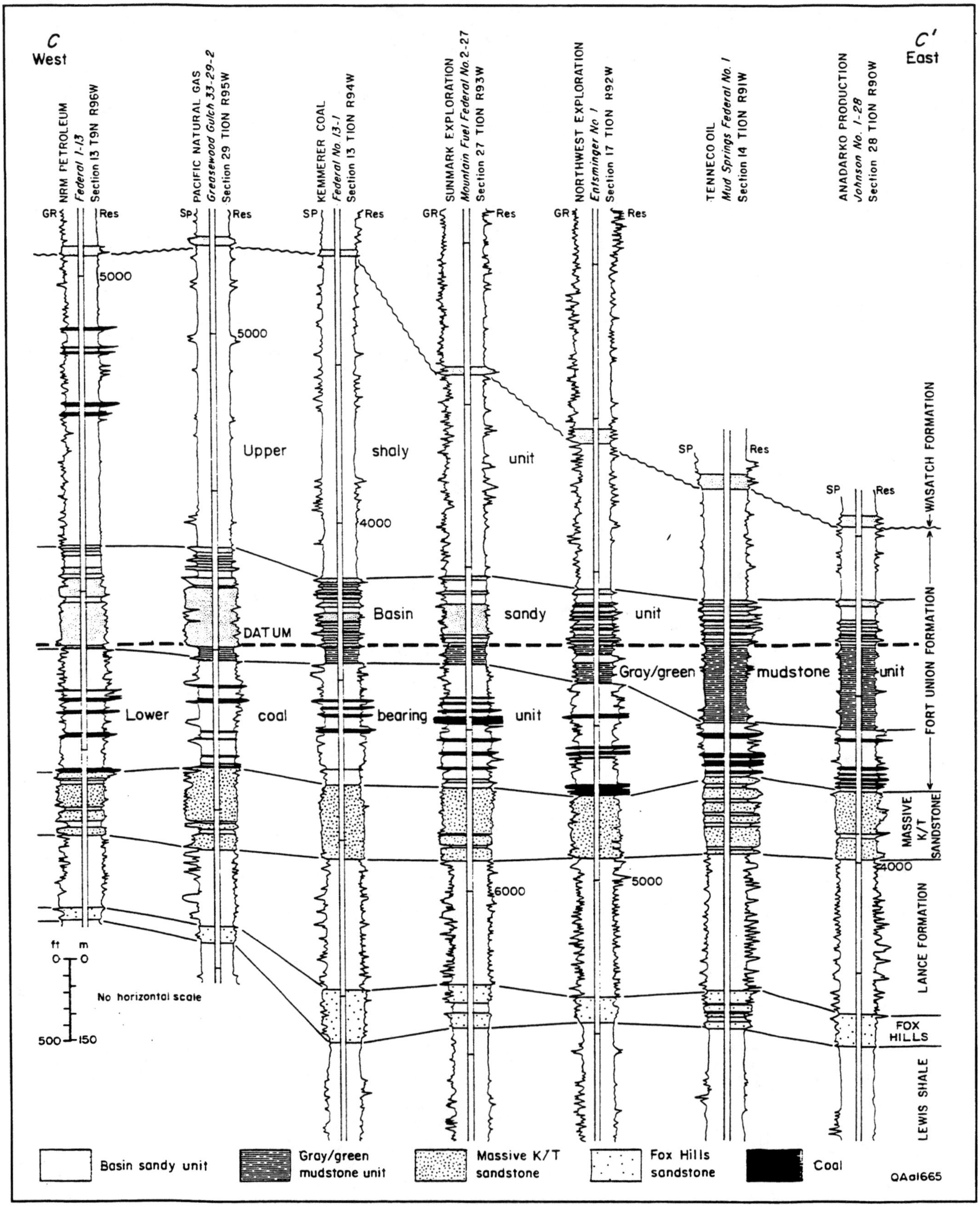

Figure 51. West-east stratigraphic cross section C–C′ through T10N of the Paleocene Fort Union Formation, Sand Wash Basin, illustrating operationally defined stratigraphic units and coal occurrence of the lower coal-bearing unit. Thickest and more continuous coal beds occur in the eastern Sand Wash Basin above thickest sandstone development.

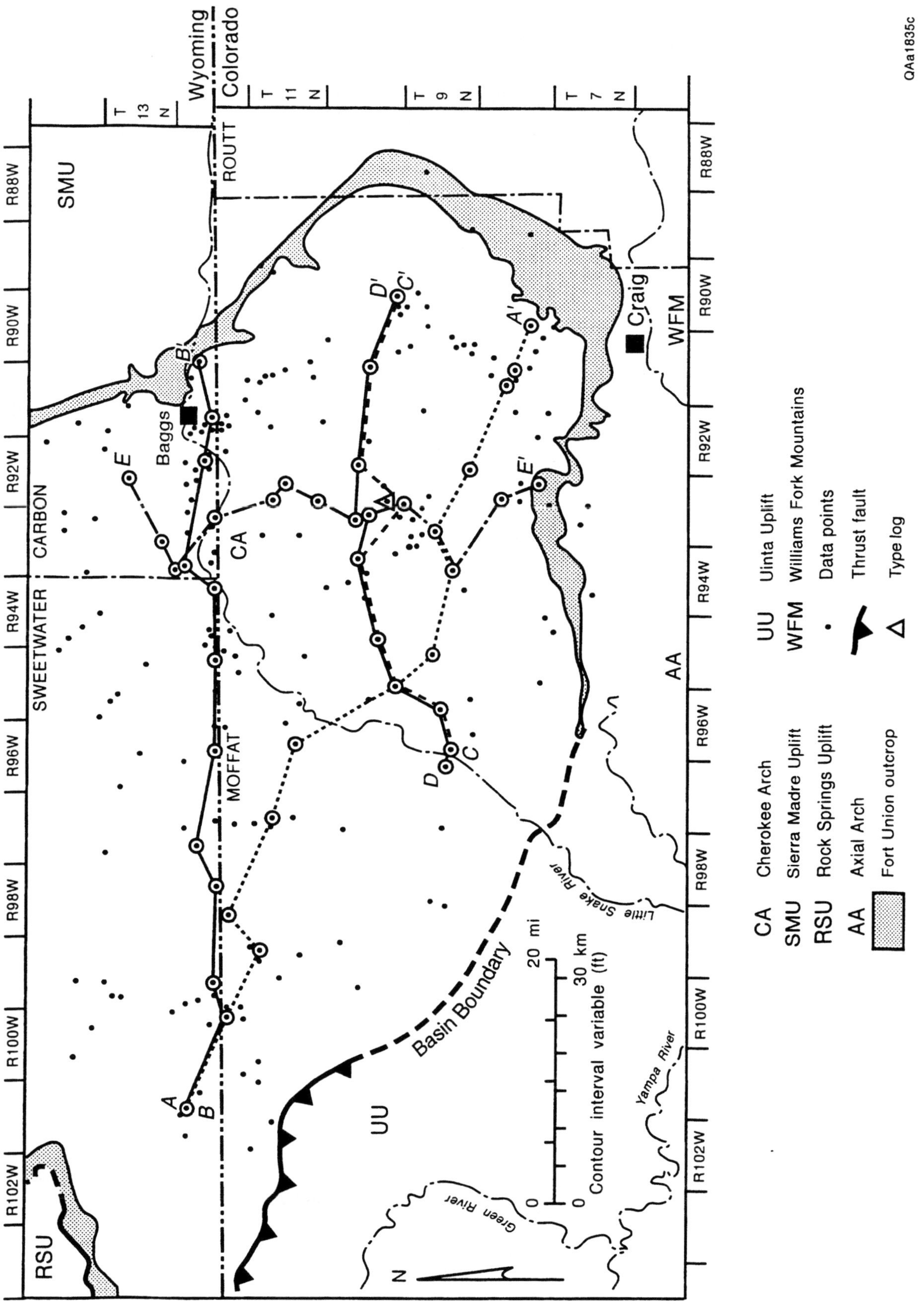

Figure 52. Location of stratigraphic cross sections A–A′ to E–E′ (figs. 49, 50, 51, 57, and 58) in the Paleocene Fort Union Formation, Sand Wash Basin.

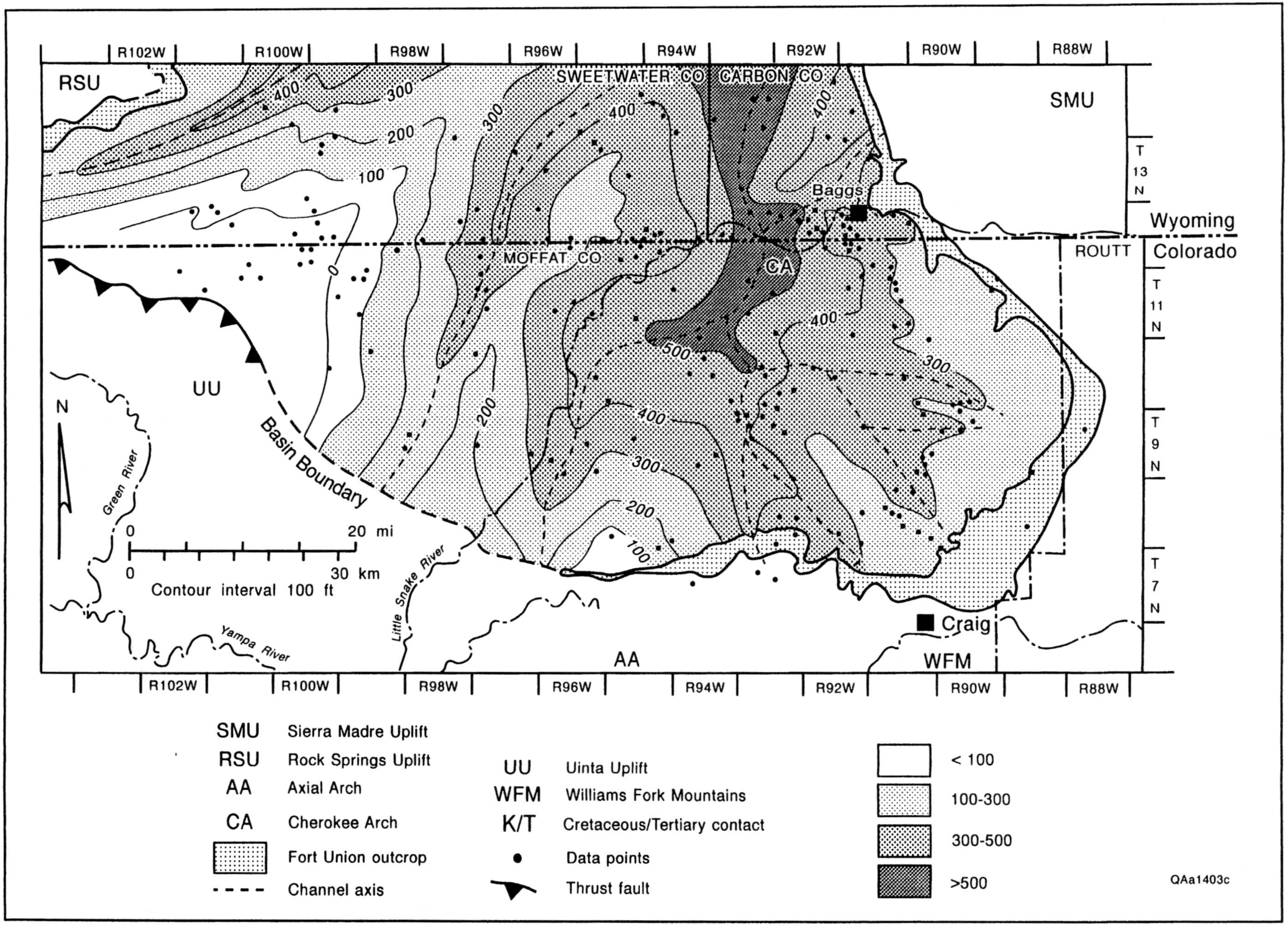

Figure 53. Net-sandstone-thickness map of the massive Cretaceous/Tertiary (K/T) sandstone unit, Sand Wash Basin. Thickest sandstone development (>500 ft [>152 m]) occurs along a south-north-oriented fluvial system between R92W and R94W.

numerous northeast- to northwest-trending tributary streams that merge with a south-north-oriented axial channel complex centered on R93W. Paleocurrent measurements taken from outcrop (Beaumont, 1979) are consistent with the subsurface data and reveal a high dispersion of current directions from 280° to 80°, with a mean vector that is oriented north-northwestward (351°). The fluvial system thickens from about 300 ft (~91.4 m) in the east (T12N, R91W) to 500 ft in T12N, R93W and then dramatically thins to less than 100 ft (<30.5 m) in the west (T12N, R100W, R101W) (fig. 53). At its thickest development between R92W and R94W, the unit is composed of laterally extensive, multistoried, amalgamated, sandstone sequences as much as 200 ft (60.9 m) thick, with individual sandstone bodies as much as 50 ft (15.2 m) thick (figs. 48–50). The sandstone sequences are separated by mudstones 1 to 50 ft (0.3 to 15.2 m) thick. The westward thinning is the result of erosion at the Cretaceous/Tertiary unconformity, as well as lateral facies changes into the Fort Union Formation. Hettinger and Kirschbaum (1991) similarly interpreted westward thinning of the massive K/T sandstone in the Washakie Basin as a result of erosion at the Cretaceous/Tertiary unconformity, and lateral facies change into the Fort Union Formation.

The massive K/T sandstone unit is further subdivided into lower and upper zones based on the presence of a regional unconformity. The lower zone is, in part, laterally equivalent to some of the sandstone in the upper part of the Lance Formation (Hettinger and others, 1991). The lower zone is separated from the upper zone by an erosional surface, which is usually depicted in outcrop (east and west Sand Wash Basin) by a distinct conglomerate horizon, representing the unconformity between Cretaceous and Tertiary rocks. Palynology indicates that the lower zone is Upper Cretaceous and the upper zone is Paleocene (Hettinger and others, 1991). The upper (Paleocene) sandstone overlying the basal conglomerate horizon is as much as 220 ft (67 m) thick in the eastern Sand Wash Basin and consists of multistoried blocky sandstone bodies (figs. 49–51). Interbedded with the sandstone bodies are a few thin (<10 ft [<3.05 m] thick) shales. To the west the upper zone is thinner and contains sandstones that intertongue with shale and coal beds that are equivalent to the basal part of the lower coal-bearing unit of the Fort Union Formation. The massive K/T sandstone has no significant coal beds.

Fort Union Formation

The operational base of the Paleocene Fort Union Formation is placed on top of the massive K/T sandstone unit. The Fort Union Formation is operationally subdivided into the lower coal-bearing unit, the gray-green mudstone unit, the basin sandy unit, and the upper shaly unit (fig. 48). In the east and west parts of the Sand Wash Basin, the lower coal-bearing unit is overlain by the noncoal gray-green mudstone unit, the basin sandy unit, and the upper shaly unit, but only the basin sandy unit and upper shaly unit in the center (between R98W and R101W) (figs. 49–51). Regionally, the Fort Union Formation, as defined here, thickens to the west from 1,300 ft (396 m) (T8N, R91W) to between 2,600 and 3,000 ft (792 and 914 m) (T12N, R96W) and then thins to between 1,600 and 2,000 ft (488 and 609.6 m) (T12N, R101W) (fig. 50). Thickness of the Fort Union Formation reflects its depositional setting and/or periods of nondeposition and erosion along the Eocene–Paleocene (Wasatch Formation–Fort Union Formation) unconformity. In the northeastern part of the Sand Wash Basin (T12N, R91W), the approximate depth to the base of the Fort Union Formation coal seams is less than 2,000 ft (<609.6 m) (Tyler, this vol.). These coal beds are as much as 50 ft (15.2 m) thick and combine for a typical net-coal thickness of 80 ft (24.4 m) (in as many as nine coal beds). Coal beds are laterally continuous and correlatable into the eastern part of the basin for up to roughly 18 mi (29 km) between R90W and R95W (figs. 49–51). Between R95W and R98W, the Fort Union Formation coal seams are commonly as much as 10 ft (3.05 m) thick, less continuous, combine for net-coal thicknesses of less than 30 ft (<9.1 m), and are more numerous (as many as 11 coal seams). Approximate depth to the base of the Fort Union Formation is >8,000 ft (>2,438 m) in the deepest part of the basin (Tyler and Tremain, this vol). West of R98W, the Fort Union coal beds are thicker (≤20 ft [≤6.1 m]) and shallower (>2,500 ft [>762 m]); the Fort Union Formation crops out along the Rock Springs Uplift. The Fort Union Formation is the major Paleocene coal and coalbed methane target in the Sand Wash Basin.

Wasatch Formation

The main body of the Wasatch Formation overlies the upper shaly unit of the Fort Union Formation (figs. 49–51) and can exceed 2,000 ft (609.6 m) in thickness in the basin center but thins markedly to less than 500 ft (<152 m) on the eastern margins of the Sand Wash Basin. In the Washakie Basin the contact between the Wasatch Formation and the underlying upper shaly unit of the Fort Union Formation is marked by an erosional surface and is interpreted to be a disconformity (Hettinger and others, 1991). Hettinger and others (1991) placed the base of the Wasatch Formation below a basal conglomeratic zone in surface and subsurface studies. Where the conglomerate zone is absent, the base is placed by utilizing descriptions of drill-hole readings from the American Stratigraphic Company drill hole descriptions and/or on the first occurrence of a varicolored mudstone (Hettinger and others, 1991). These descriptions are generally consistent with field observations by Hettinger and others (1991) regarding thickness, grain size, and lithology of the Wasatch Formation. Seismic lines provided to the Bureau of Economic Geology by Union Pacific Resources also helped identify the Fort Union/Wasatch unconformity in the Sand Wash Basin. On geophysical well logs in the western Sand Wash Basin, the base of the Wasatch Formation is characterized by sharp spontaneous potential, gamma ray, and resistivity responses associated with influx of fresh water along channel-fill sandstones. In outcrop near Baggs, the contact between these formations is placed at the first occurrence of coarse-grained or conglomeratic sandstone overlain by variegated mudstone (Hettinger and others, 1991). West of Baggs in R91W the contact is placed where a coarse-grained sandstone overlies coal beds of the Cherokee coal zone. Depositionally, the Wasatch Formation consists of conglomeratic lacustrine fan-delta deposits that grade laterally into fluvial sandstones, floodplain and lacustrine shales, and minor coal-bearing floodplain deposits (Roehler, 1965; Sklenar and Anderson, 1985). As operationally defined herein, the Wasatch Formation contains no significant coal in the Sand Wash Basin and therefore is not a coalbed methane target.

Sandstone and Coal Occurrence of the Fort Union Formation

Sandstone and coal occurrence maps of the Fort Union Formation were prepared to contrast their distribution in the Sand Wash Basin. Sandstone and coal beds were identified from the analysis of geophysical well log signatures. On geophysical well logs, coal was identified by low density, high neutron and density porosities, low sonic velocity, and/or low neutron count. On geophysical logs, the thickness of a bed is commonly measured halfway between the shale baseline and the peak corresponding to that bed. On the bulk density log, coal-seam thickness was measured at a density of approximately 1.80 g/cm^3. We recorded the thickness of coal seams thicker than 2 ft (0.61 m); partings thinner than 2 ft (0.61 m) within thick coal seams were included as coal because of the limits of resolution of the geophysical logs. When density and sonic logs were unavailable, very high resistivities along with either a low gamma or a shalelike resistivity response were used to operationally define the coal packages.

Lower Coal-Bearing Unit

The lower coal-bearing unit thickens to the west from 500 ft (152 m) (T8N, R91W) to 900 ft (274 m) (T12N, R98W) and then thins to about 300 ft (~91.4 m) (T12N, R101W) (figs. 49–51); it consists mainly of repetitive, upward-fining, upward-coarsening, and/or blocky, amalgamated sandstone sequences that average about 140 ft (~42.7 m) thick, interbedded with thin shales (<80 ft [<24.4 m] thick) and thick coal (<50 ft [<15.2 m] thick). The net-sandstone-thickness map indicates a northward-flowing fluvial system with its depositional axis centered in R93W (fig. 54). In R93W net-sandstone thicknesses are greater than 500 ft (>152 m), thinning east and west to less than 200 ft (<60.9 m) in T7N–T11N, R91W and T7N–T10N, R96W, respectively. The sandstone-percent and the maximum-sandstone isopach maps indicate similar trends. The thickest sand development (>70% and >120 ft [>36.6 m] [figs. 55 and 56, respectively]) occurs in a north-trending belt centered on R93W, confirming the position of the axis of the north-flowing trunk-stream system. Two smaller tributaries flowed east from the Uinta Uplift, and several tributaries flowed west from the Sierra Madre–Park Uplift.

In south-north- and east-west-oriented cross sections, two major south-north-trending coal packages accumulated above and on the flanks of main trunk systems between R91W and R94W (figs. 57 and 58). The lower coal package (package 1) rests on, or up to 100 ft (30.6 m) above, the thickest development of the massive K/T sandstone (figs. 56 and 57). Coal-occurrence mapping suggests that this south-north-trending coal package is thickest along R91W and extends from the southern Sand Wash Basin into the Washakie Basin (fig. 59). The thickest individual coal bed in the package is as much as 35 ft (10.7 m) thick. The upper coal package (package 2) lies about 300 ft (~91.4 m) above the lower package and is similarly oriented south-north, extending into the Washakie Basin (figs. 57–59). Individual coal beds in the upper package are ≤50 ft (≤15.2 m) thick and are centered on the boundary between R92W and R93W (fig. 59). The thicker coals in the upper package overlie or flank thick channel-fill sandstone sequences.

Coal-occurrence mapping of the individual coal beds throughout the Sand Wash Basin indicates ranges in thickness from less than 10 ft (<3.05 m) to as much as 50 ft (15.2 m) (fig. 59). Maximum individual coal bed thicknesses are greatest (>40 ft [>12.2 m]) along R93W, following the same trend as the net-sandstone thickness (fig. 54). Coal beds dramatically thin (<10 ft [<3.05 m]) east and west of this central trunk system. Net-coal thicknesses follow similar trends, with a maximum net-coal thickness of greater than 80 ft (>24.4 m) along the

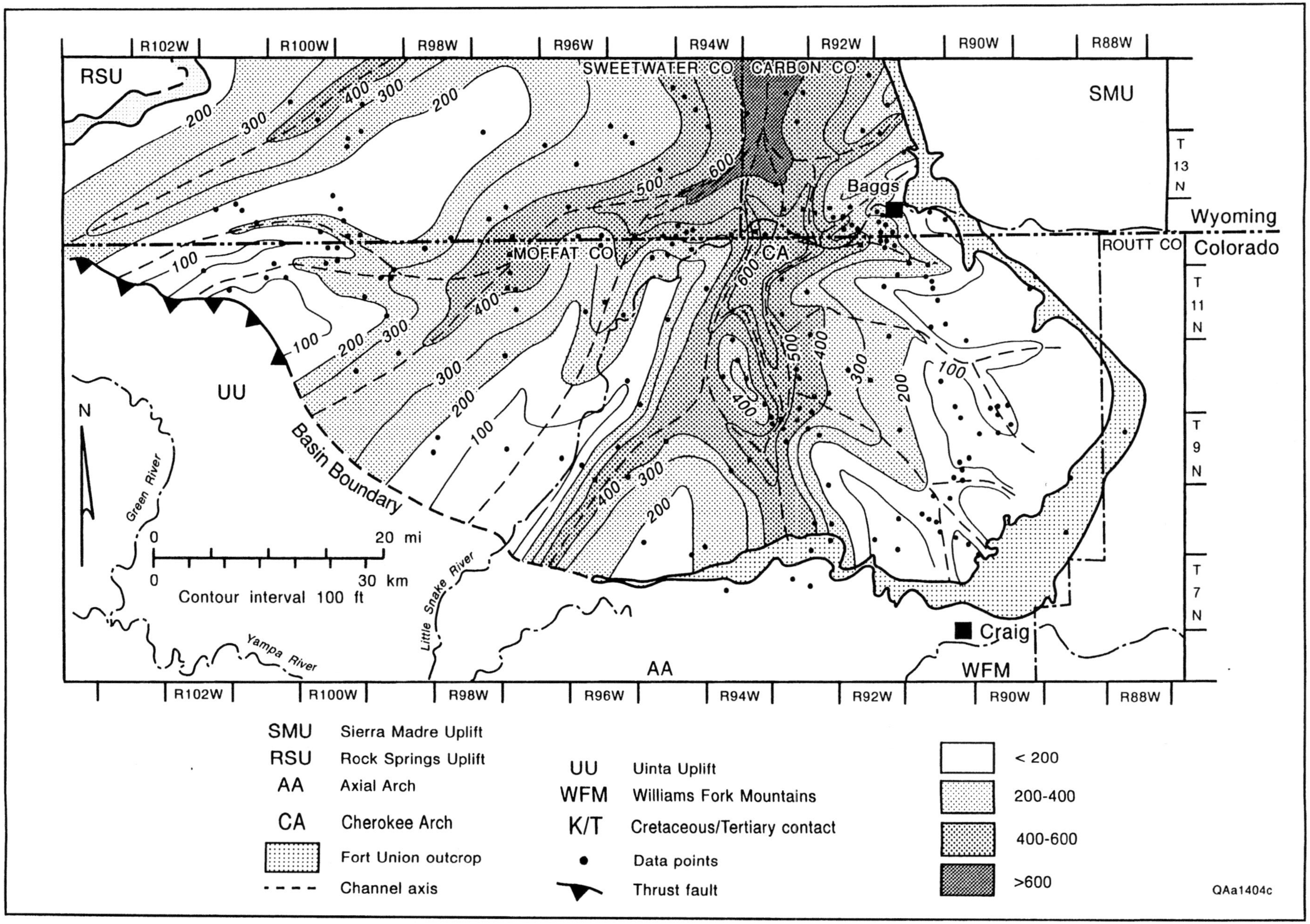

Figure 54. Net-sandstone-thickness map of the lower coal-bearing unit, Fort Union Formation, Sand Wash Basin. Thickest sandstone development (>600 ft [>182.8 m]) occurs along a south-north-oriented fluvial system between R92W and R94W. Smaller tributaries flow both east and west toward the depositional axis centered on R93W.

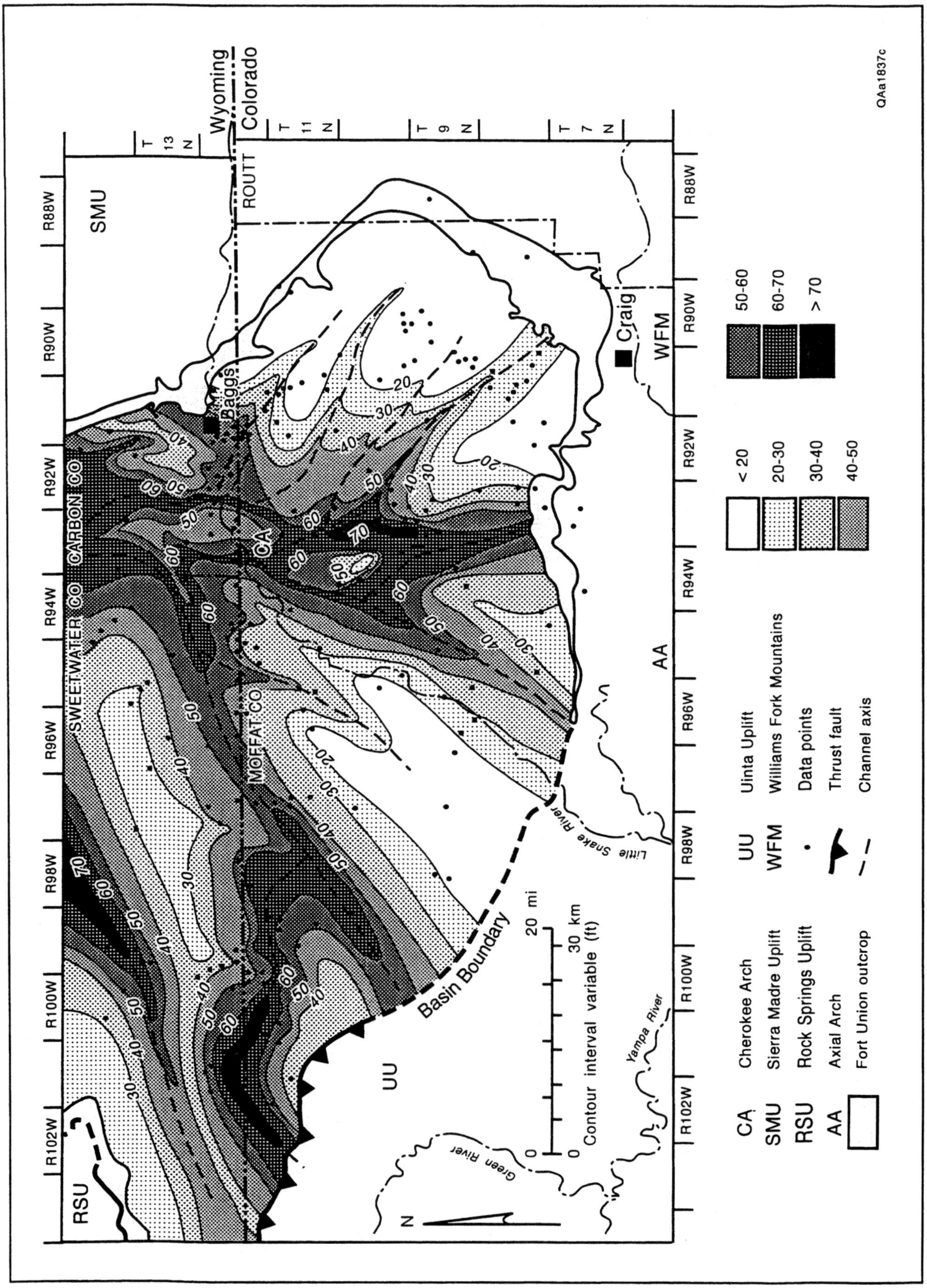

Figure 55. Net-sandstone-percentage map of the lower coal-bearing unit, Fort Union Formation, Sand Wash Basin. Thickest sandstone development (>60%) occurs along a south-north-oriented fluvial system between R92W and R94W.

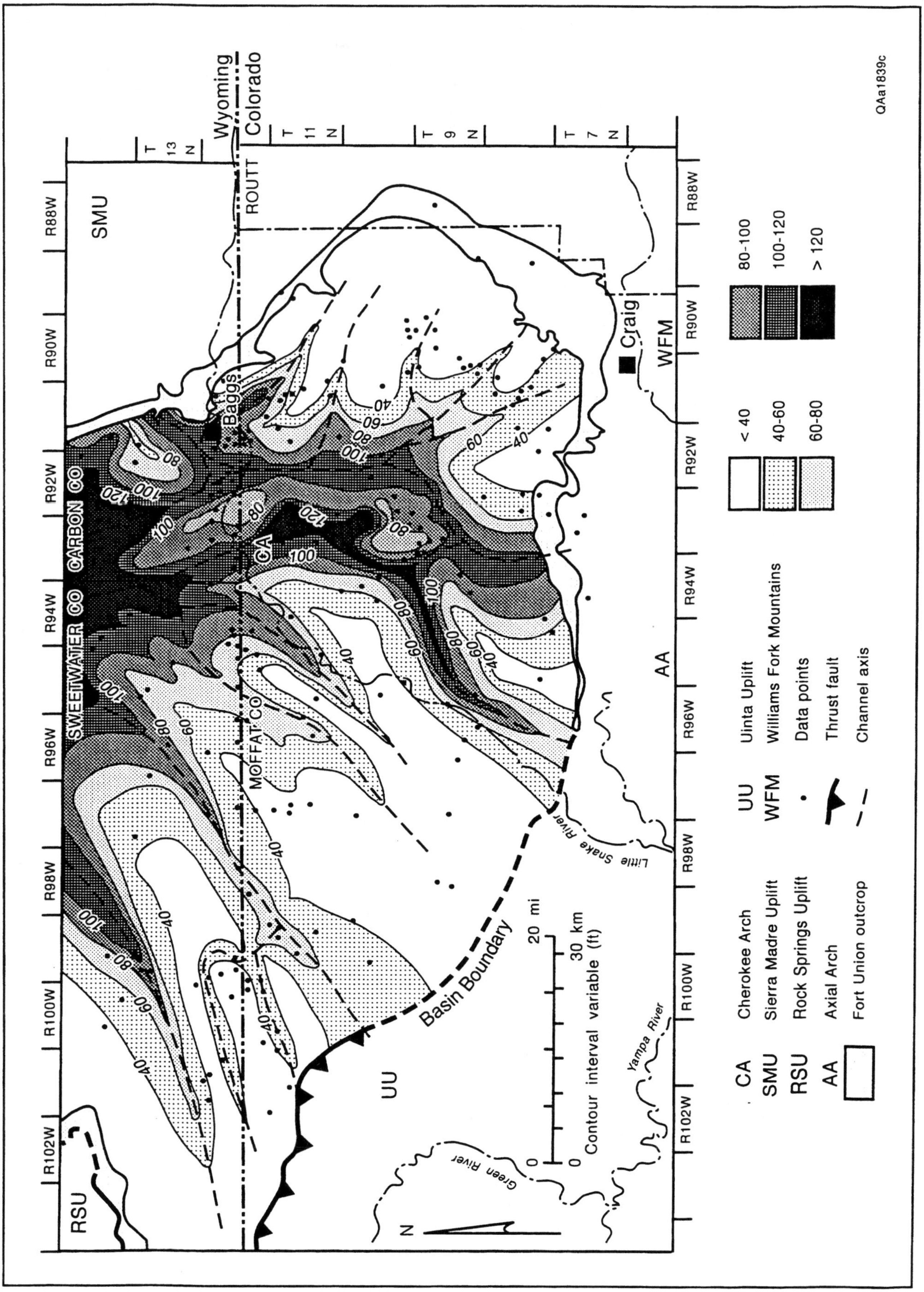

Figure 56. Maximum-sandstone-thickness map of the lower coal-bearing unit, Fort Union Formation, Sand Wash Basin. Thickest individual sandstone development (>120 ft [>36.6 m]) occurs along a south-north-oriented fluvial system between R92W and R94W.

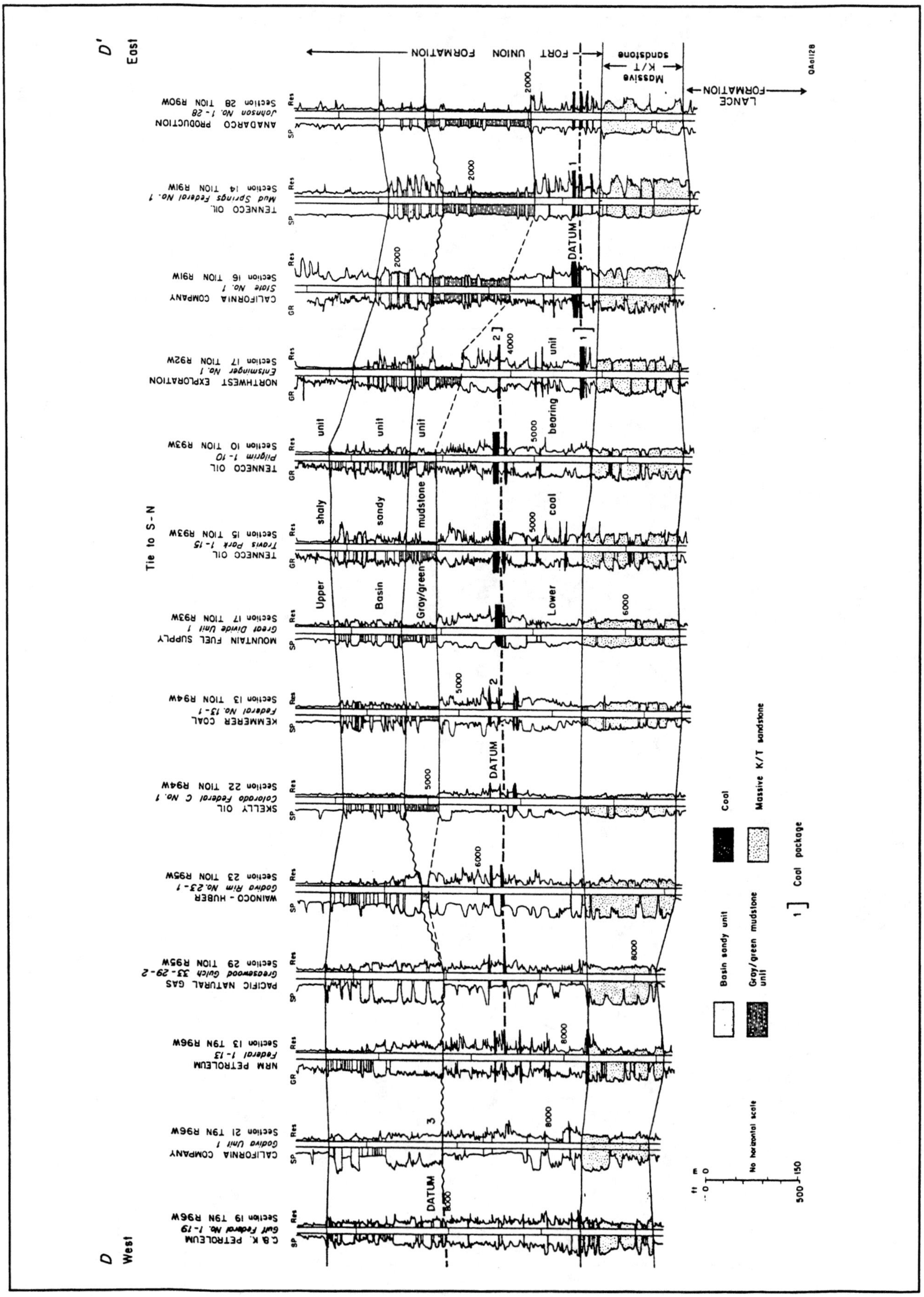

Figure 57. Detailed west-east stratigraphic cross section D–D′ through T12N of the lower coal-bearing unit, Paleocene Fort Union Formation, Sand Wash Basin, illustrating occurrence of coal packages 1 and 2.

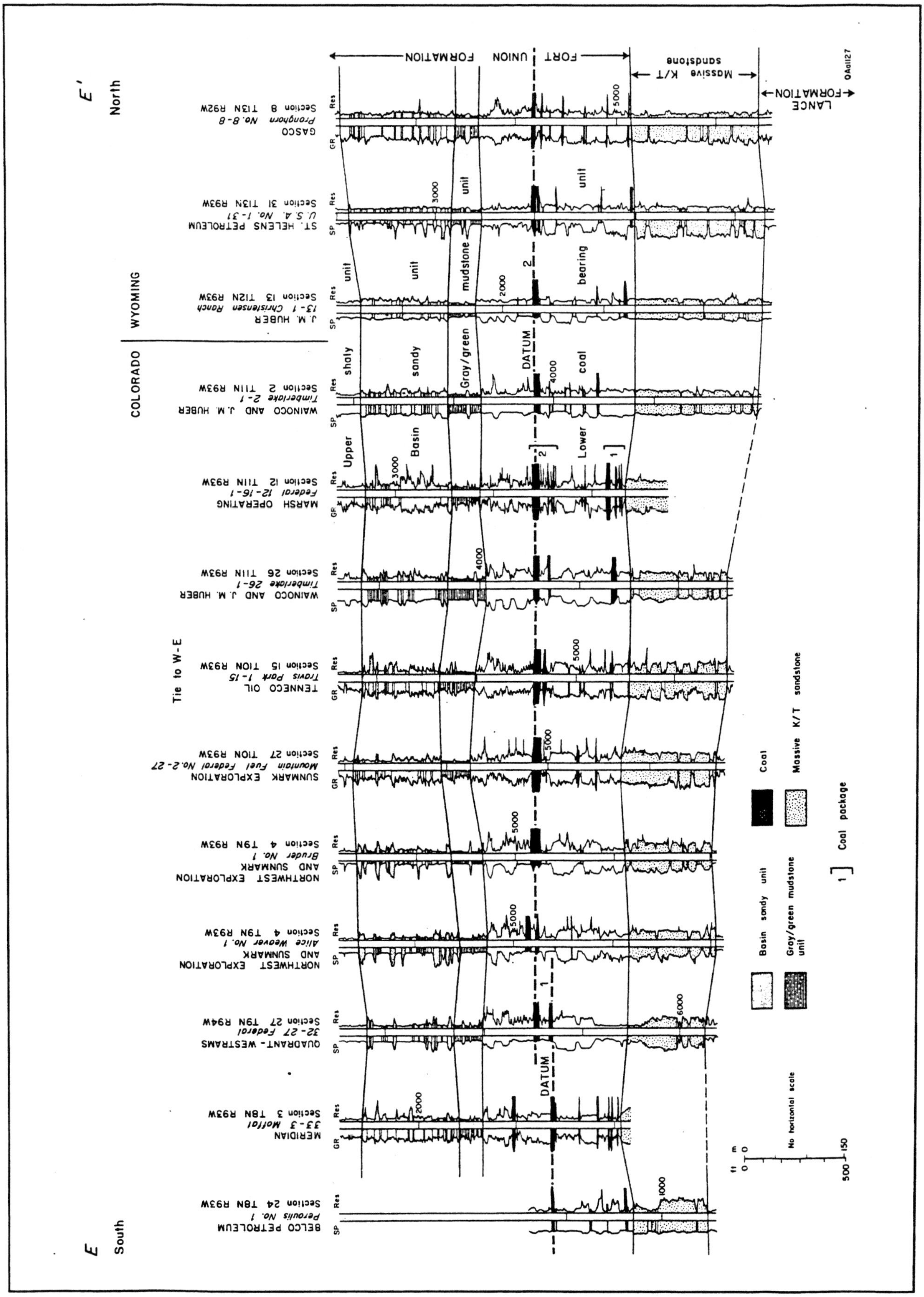

Figure 58. Detailed north-south stratigraphic cross section E–E′ between R92W and R93W of the lower coal-bearing unit, Paleocene Fort Union Formation, Sand Wash Basin, illustrating occurrence of coal packages 1 and 2.

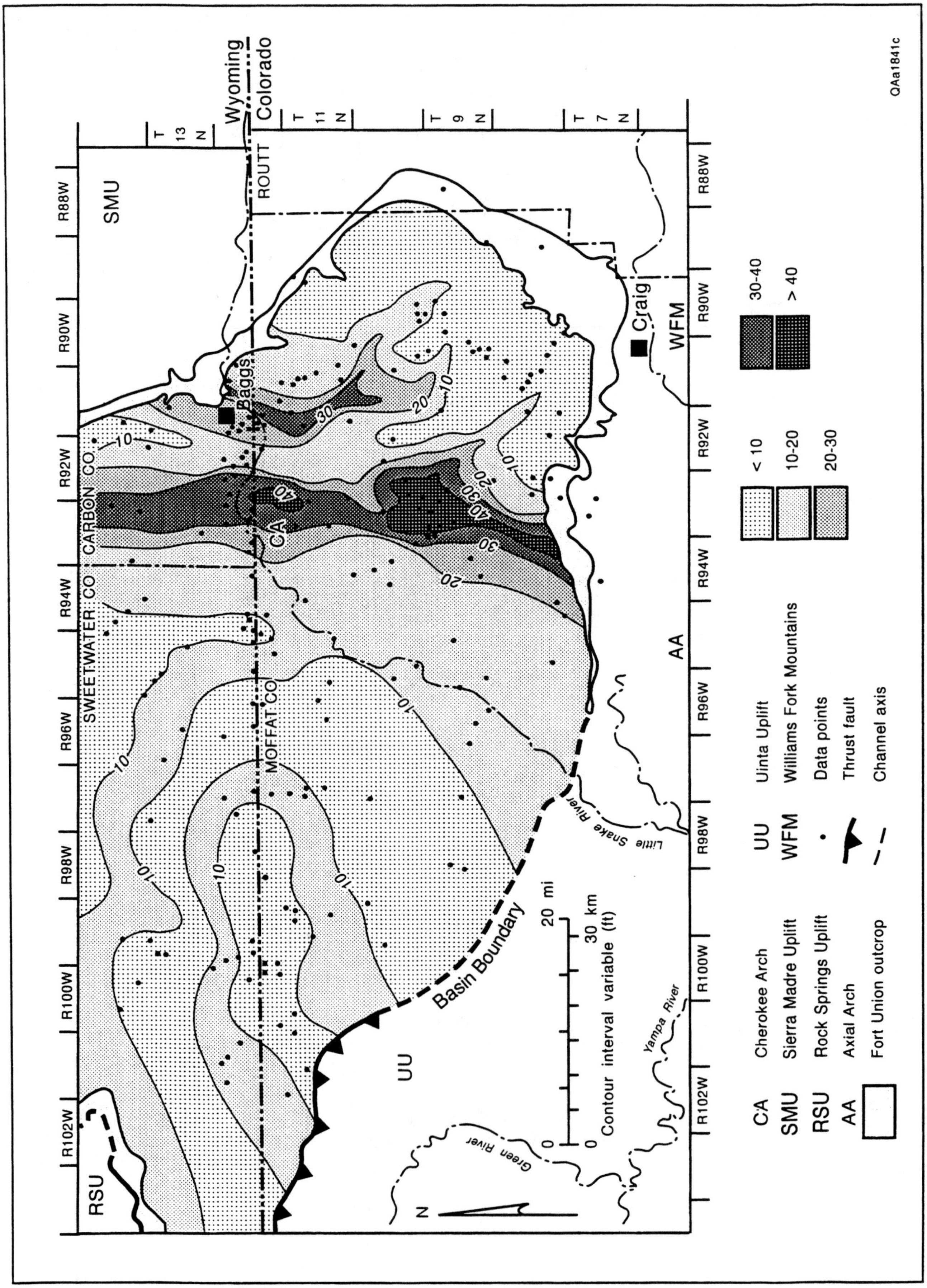

Figure 59. Maximum-coal-thickness map of the lower coal-bearing unit, Fort Union Formation, Sand Wash Basin. Thickest individual coal development (>40 ft [>12.2 m]) occurs above and alongside a south-north-oriented fluvial system between R92W and R94W.

boundary of R92W and R93W and in the Baggs area (T12N R91W) (fig. 60). Net-coal thicknesses are less than 20 ft (<6.1 m), west and southeast of the main trend. The coal-isopleth map for the lower coal-bearing interval displays patterns similar to the net-coal-thickness map. The number of seams is highest in north-oriented trends (fig. 61). Two to 12 coal beds may be present in the Fort Union Formation in various parts of the basin, the greater number being located close to Baggs and between T10N and T11N, R95W. The number of coal beds decreases to less than two east of R90W and to less than four west of R96W.

In summary, the lower coal-bearing unit represents coarse clastic deposition from several tectonically active source areas to the south, southwest, and southeast of the Sand Wash Basin. Sediment influx was predominantly from the Sawatch Range, minor input coming from the Uinta and Sierra Madre–Park Uplifts. The distribution of coal packages is in part controlled by the occurrence of the thick fluvial channel-fill sandstone sequences. To preserve a vegetation mat as peat, coal beds require a stable platform to mitigate subsidence and to provide the hydrologic conditions necessary for coal development. Syntectonic sedimentation and/or major upstream avulsion of the fluvial channel complex shuts off the coarse clastic sediment supply, forming large interchannel depressions, which are isolated from clastic input. Greatest coal accumulation occurs above and on the flanks of these thick channel-fill sandstone sequences in the Sand Wash Basin, confirming a direct relation between the position of the streams that deposited the sandstone and the location of coal. The fluvial axes of the massive K/T sandstone unit and the lower coal-bearing unit provided the stable platform and the hydrologic conditions necessary for the development of thick coal packages.

Gray-Green Mudstone Unit

The non-coal-bearing, gray-green mudstone unit (gray-green mudstone of Honey and Hettinger, 1989; stagnant lake of Colson, 1969) is present only in the western and eastern parts of the Sand Wash Basin. The gray-green mudstone is recognized on geophysical logs by its distinctive low-resistivity response and corresponds to the lower part of Beaumont's (1979) "upper shaly zone" of the Fort Union Formation, whereas McDonald (1975) included it within the Eocene Wasatch Formation. The gray-green mudstone unit forms two north-oriented belts centered on Baggs (R91W) and R101W, respectively (fig. 62). The unit was eroded in the center of the basin and north into the Washakie Basin, upon deposition of the basin sandy unit, which disconformably overlies the gray-green mudstone. The gray-green mudstone unit is about 500 ft (~152 m) thick in a northerly oriented belt centered on R91W, but it thins to <10 ft (<3.05 m) between R96W and R98W. On the west margin of the basin the unit averages 100 to 200 ft (30.6 to 60.9 m) thick (between T12N, R98W and T12N, R101W). The gray-green mudstone unit is probably lacustrine and/or floodplain in origin, reflecting tectonic quiescence, subsidence, and nondeposition of coarse clastics. The gray-green mudstone is devoid of coal and is not a coalbed methane target.

Basin Sandy Unit

The non-coal-bearing, basin sandy unit (basin sandy interval of Colson [1969], or portion of the unnamed upper Paleocene unit of Hettinger and others [1991]) overlies the gray-green mudstone unit in the Sand Wash Basin, except between R95W and R98W, where it disconformably overlies the lower coal-bearing unit. The depositional axis of the basin sandy unit is oriented south-north and thickens westward from 100 ft (30.6 m) near R91W to about 500 ft (~152 m) in R96W and R97W (fig. 63); it appears to be restricted to the central parts of the Sand Wash Basin. In the basin center, the basin sandy unit has thick (140 ft [42.7 m]), laterally extensive sandstone bodies interbedded with thin mudstones (<20 ft [<6.1 m] thick). Within the basin sandy interval, coarse-grained to conglomeratic sandstones, with an abundance of feldspar and chert, intertongue northward with medium-grained sandstones (Colson, 1969). The basin sandy interval is a south-north fluvial depositional system composed of multistoried channel-fill sandstones that is devoid of coal and is not a coalbed methane target.

Upper Shaly Unit

The upper shaly unit occurs above the basin sandy unit and includes the Cherokee coal zone along the west and west-central margins of the Sand Wash Basin. The upper shaly unit thickens to the west from 300 ft (91.4 m) near R91W to over 1,300 ft (396 m) (T12N, R96W) and then thins to about 200 to 300 ft (~60.9 to 91.4 m) in T12N, R101W (fig. 63). This thinning is due to erosion associated with a major unconformity at the base of the Eocene Wasatch Formation; erosion has removed much of the unit on the east and west margins of the basin. The Cherokee coal zone, constituting the upper 250 to 450 ft (76.2 to 137 m) of the upper shaly unit, occurs only in the west part of the Sand Wash Basin (west of R96W). This zone has previously been included in the Wasatch Formation (Smith and others, 1972); however, palynologic data show the Cherokee

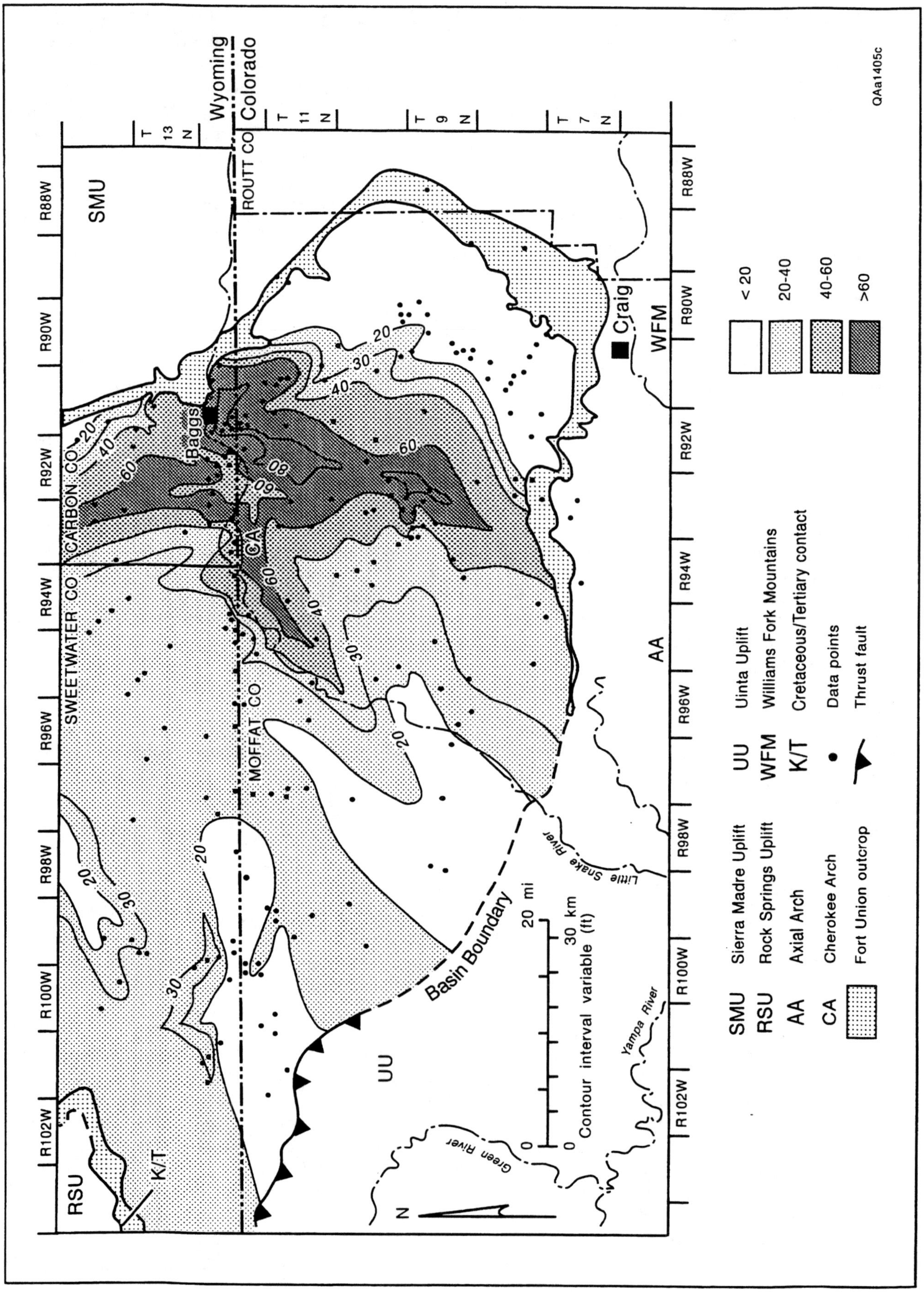

Figure 60. Net-coal-thickness map of the lower coal-bearing unit, Fort Union Formation, Sand Wash Basin. Thickest coal bed development (>60 ft [>18.3 m]) occurs above and alongside a south-north oriented fluvial system between R92W and R94W.

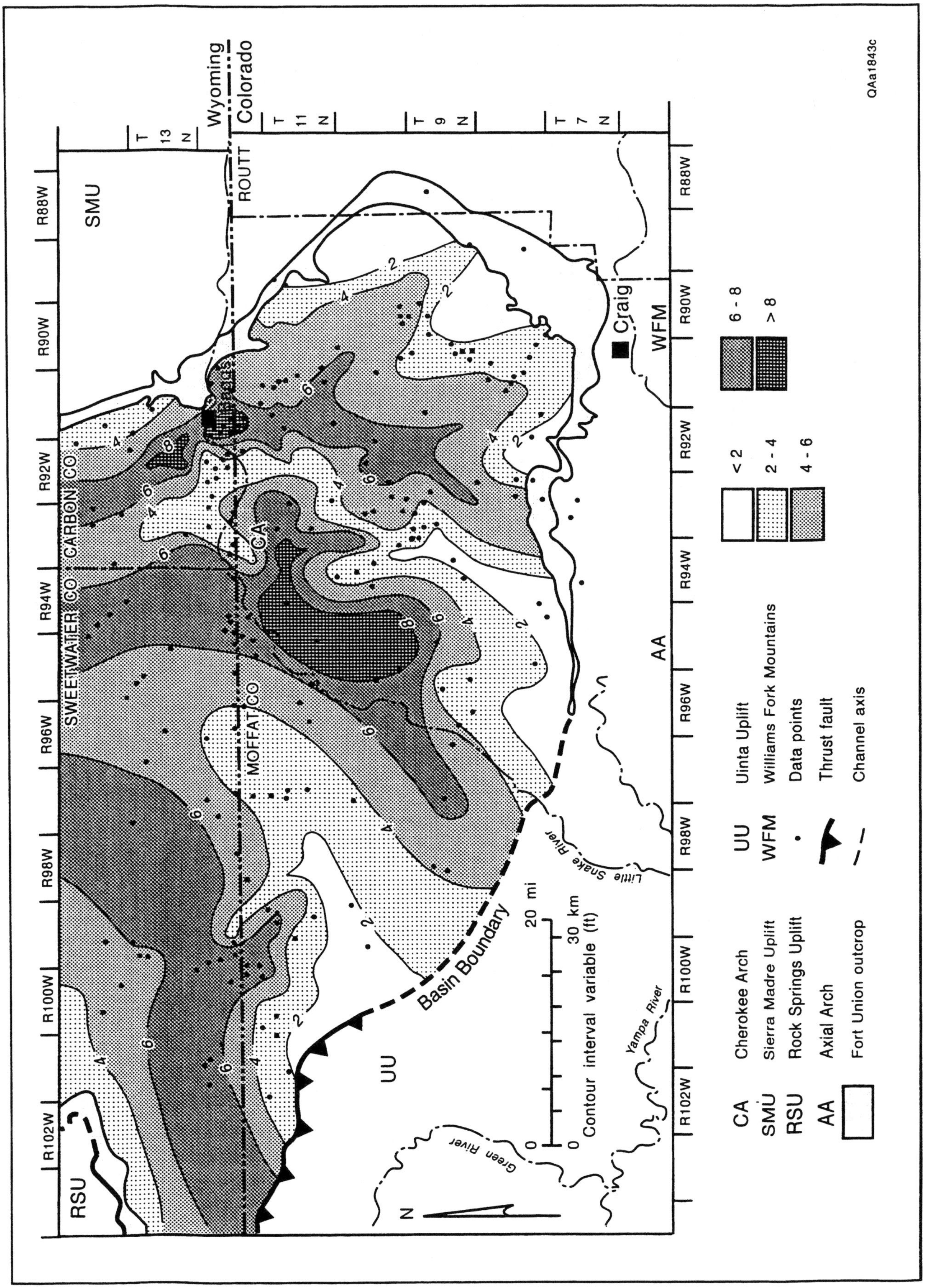

Figure 61. Coal-isopleth map of the lower coal-bearing unit, Fort Union Formation, Sand Wash Basin. Greatest number of coal beds (>8) occurs above and alongside a south-north-oriented fluvial system between R92W and R94W.

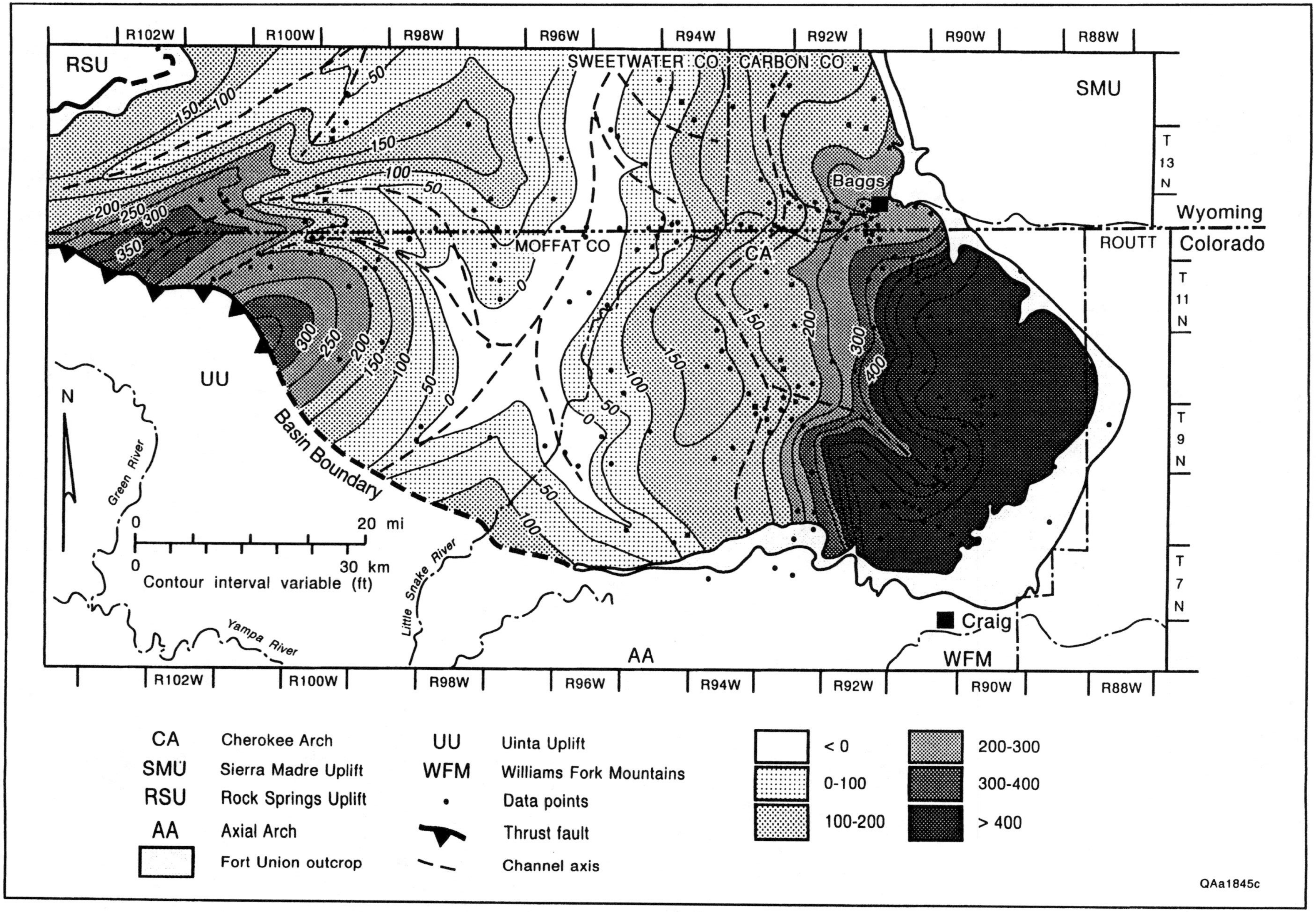

Figure 62. Net-mudstone-thickness map of the gray-green mudstone unit, Fort Union Formation, Sand Wash Basin. Thickest mudstone development (>400 ft [>121.9 m]) occurs east of the south-north-oriented fluvial system between R92W and R94W, adjacent to the Sierra Madre–Park Uplift. Structural and depositional axes have migrated to the west (R96).

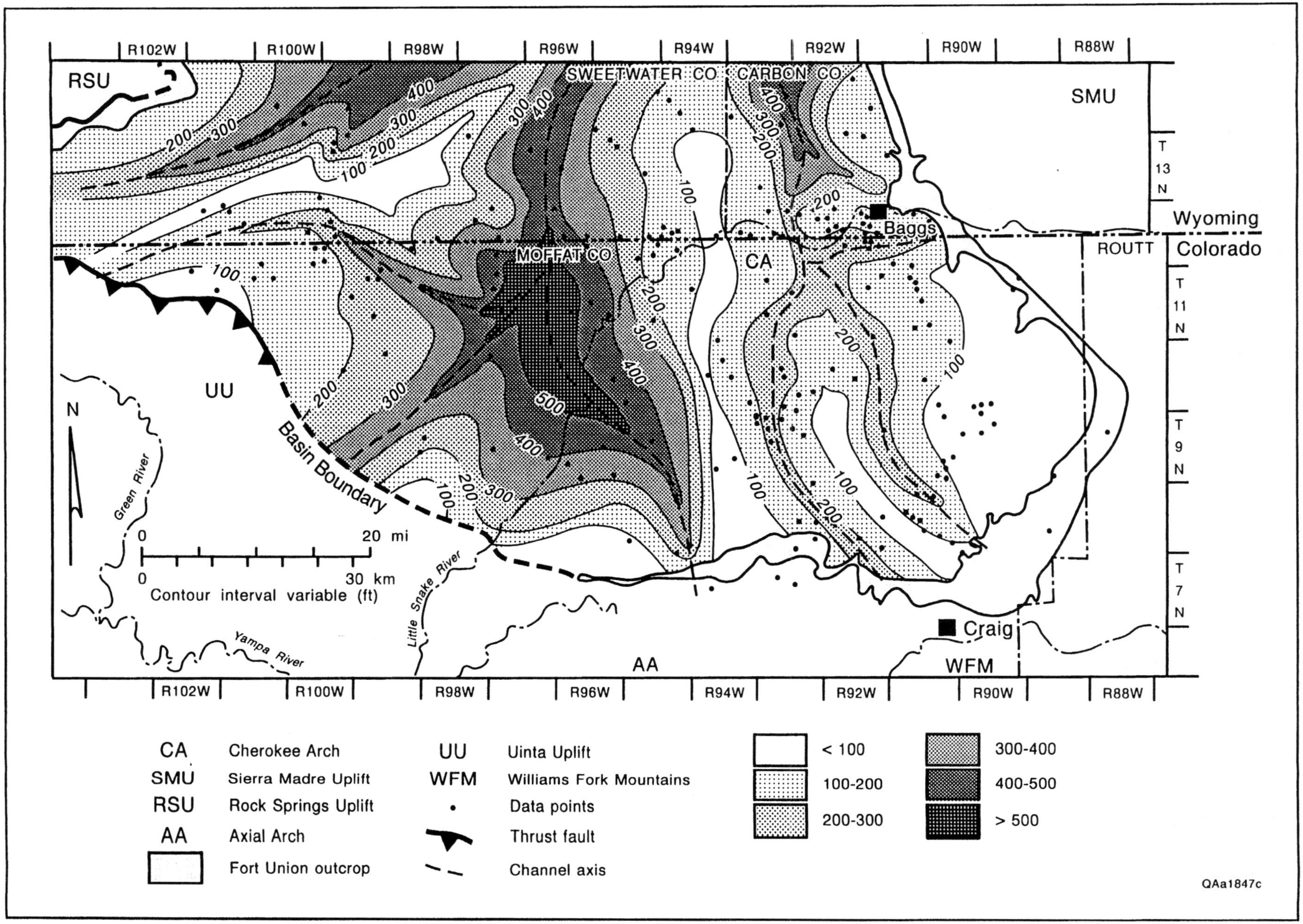

Figure 63. Net-sandstone-thickness map of the basin sandy unit, Fort Union Formation, Sand Wash Basin. Thickest sandstone development (>500 ft [>152 m]) occurs along a south-north-oriented fluvial system centered on R96W.

coals to be Paleocene in age (Hettinger and others, 1991). The Cherokee coal beds are lenticular and thin (3 to 10 ft [0.91 to 3.05 m] thick) and pinch out to the east because of lateral facies changes into the basin. They also are disconformably removed by the Wasatch Formation downcutting to the east of R95W. The upper shaly unit along the western portion of the basin is predominantly a mixed-load fluvial system; eastward, the system gives way to lower energy suspended-load, channel-margin and/or floodplain depositional environments, which contain some sediments of a recurrent lacustrine environment. The fluvial sediment source for the unit was probably west and south of the Sand Wash Basin. The upper shaly unit has thin, shallowly buried coal beds that are minor coalbed methane targets in the Sand Wash Basin.

Geologic Controls on the Occurrence of Paleocene Fort Union Formation Coal Beds, Sand Wash Basin

Upper Cretaceous to lower Tertiary rocks in the Sand Wash Basin represent a transition from marine to continental sedimentation (Beaumont, 1979). Deposition of the marine Lewis Shale was followed by sedimentation of the nearshore marine Fox Hills Sandstone and the continental Lance Formation, massive K/T sandstone unit, and Fort Union Formation. The Fox Hills Sandstone was deposited during the final eastward advance of the Upper Cretaceous shoreline into the Western Interior Seaway and intertongues with the overlying fluvial Lance Formation. During accumulation of Lance sediments, rivers continued to flow eastward across the basin (Masters, 1961). Basement thrusting during the Laramide Orogeny resulted in the rapid emplacement of the Unita and Sierra Madre–Park Uplifts, which reoriented the drainage of the Sand Wash Basin from eastward to northward. Successive episodes of Laramide-style basement thrusting caused the tectonically induced asymmetric Sand Wash Basin subsidence to equal or exceed the rate of sedimentation.

Reorientation of the drainage pattern resulted in the development of a large intermontane river system that flowed north (fig. 64). Tweto (1975) and Beaumont (1979) have similarly suggested the Sawatch Range as the source area for the syntectonic sediments that make up the majority of the massive K/T sandstone unit. Smaller tributaries flowed east and west, contributing additional sediments to the massive K/T fluvial system (fig. 64). The Uinta Uplift to the west probably shed some sediments eastward into the Vermillion Basin area and the southwestern Sand Wash Basin. An eastern source was the Sierra Madre–Park Uplift, where early tectonic activity may have been greater than that along the Uinta Uplift. Erosion may have reached the core of the Sierra Madre–Park Uplift by early Fort Union time and coarse arkosic clastic material was deposited basinward (Colson, 1969). A high concentration of sand in R93W and northerly paleocurrent directions (Beaumont, 1979) confirm that the K/T river system entered from the south and flowed north. The broad distribution of the fluvial system indicates that tributaries ranged widely across the basin but tended to converge on their original depositional axis. The massive K/T fluvial sandstone provided the stable platform upon which peat of the lower coal-bearing unit accumulated.

The north-flowing fluvial system persisted during deposition of the overlying Fort Union Formation (fig. 64). An increase in the suspended load carried by the fluvial system resulted in the building of levees that stabilized the channel axes and allowed for the formation of extensive floodplains in which thick coal and mudstone could develop. The thick floodplain deposits underwent differential compaction, and shallow ephemeral lakes and ponds formed where the rate of compaction exceeded the rate of sediment input. Where subsidence kept pace with organic accumulation and reducing conditions prevailed, peat accumulated, resulting in the formation of coal in the *lower coal-bearing unit*. In the lower coal-bearing unit, coal and sandstone development are coincident; the thickest coals occur above or on the flanks of the thickest fluvial sandstones. Away from the fluvial axes the coal beds are split, thinner, and less continuous. The thick fluvial sandstone sequences acted as platforms for coal accumulation and conduits for ground-water flow. This is particularly true of the major north-trending channel sandstone belts, which facilitated ground-water flow basinward from a southern recharge area. Recharge was in the highlands at the basin margins and flow was basinward, down hydraulic gradient in response to the topographic gradient, for eventual discharge to topographically low areas. At these postulated sites of regional ground-water discharge, peat swamps were initiated as a result of syntectonic sedimentation and/or stream avulsion, which ultimately reduced the sediment load and allowed the peat swamps to spread across the flood plain. As a confined aquifer system, channel-fill sandstones focused discharge to initiate organic accumulation and subsequently to maintain the water table at optimal level for peat accumulation.

The *gray-green mudstone unit*, found predominantly in the western (around the edge of the Vermilion Basin) and eastern Sand Wash Basin, represents a cessation of coarse clastic sedimentation into the Sand Wash Basin. Low-energy floodplain and lacustrine depositional environments expanded to their greatest areal extent during this time. The gray-green mudstone unit was

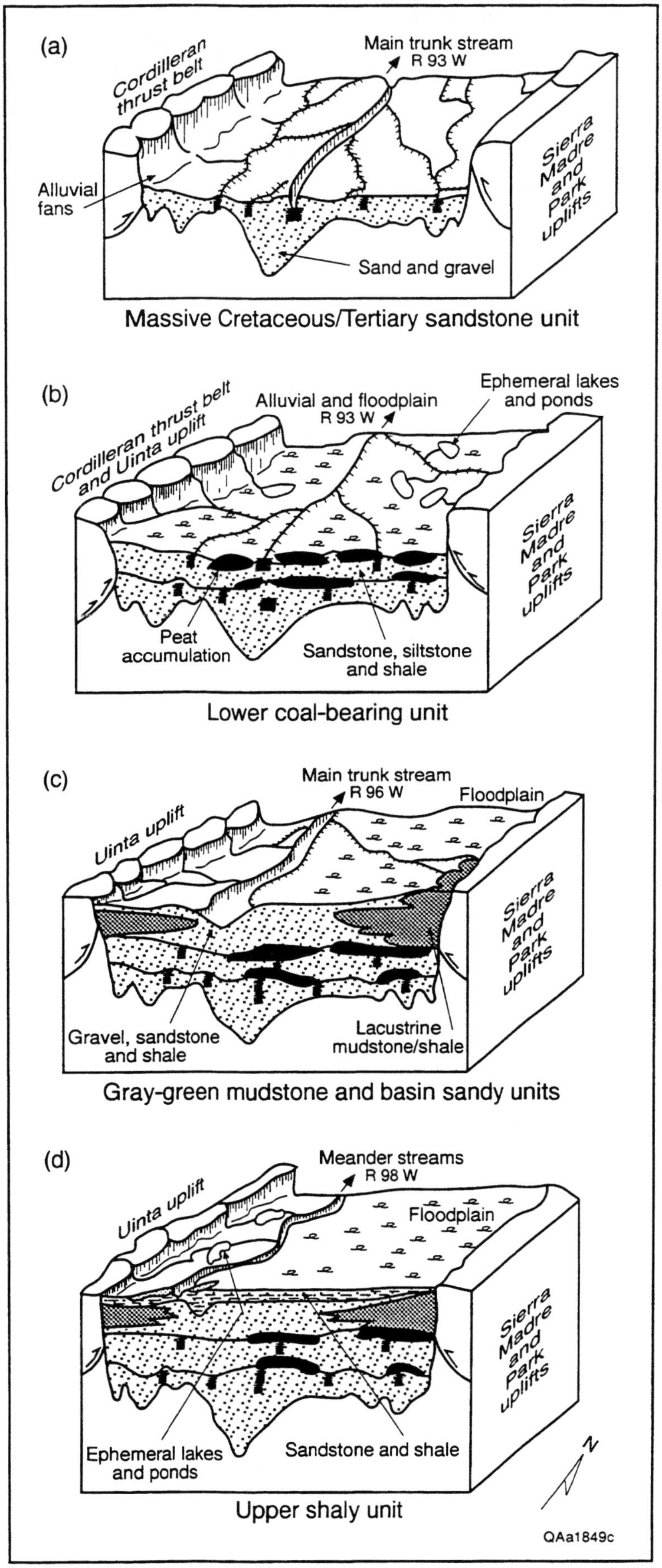

Figure 64. Block diagrams showing the stratigraphic development of the Fort Union Formation, Sand Wash Basin. (a–d) Valley aggradation stages 1–4: (a) Maximum massive K/T sandstone development occurs along a south-north-oriented fluvial system centered on R93W. (b) Lower coal-bearing unit: maximum sandstone development occurs along a south-north-oriented fluvial system; maximum coal development occurs above and alongside the fluvial system as floodplain deposits. (c) Gray-geen mudstone and basin sandy units: maximum mudstone/shale development occurs in the eastern and western Sand Wash Basin as floodplain and lacustrine deposits; maximum sandstone development occurs along a south-north-oriented fluvial system centered on R96W. (d) Upper shaly unit: maximum sandstone development occurs in the western Sand Wash Basin; the eastern Sand Wash Basin is predominantly made up of shaly floodplain and lacustrine deposits. Modified from Beaumont (1979).

subsequently eroded in the central part of the Sand Wash Basin prior to the deposition of the *basin sandy unit.* Streams flowing north, as suggested by the sandstone geometry, changed from higher sinuosity during the deposition of the gray-green mudstone unit to lower sinuosity during the deposition of the basin sandy unit. This process was probably caused by a change in base level and renewed tectonic activity, which uplifted the margins of the Sand Wash Basin, rejuvenating the streams. Streams depositing the basin sandy unit flowed predominantly in the center of the basin (R96W), in a north-trending belt, creating a stable platform for peat accumulation, similar to the platform created by the massive K/T sandstone and the lower coal-bearing units. The basin sandy unit generally is thickest and restricted to the central parts of the Sand Wash Basin, coincident with the westward-migrating structural axis of the basin.

An increase in the suspended load carried by the fluvial system brought about a further stabilization of the channel axes and again caused the formation of flood plains within the *upper shaly unit.* The fluvial sediment source for the upper shaly unit was probably west and/or south of the Vermilion and Sand Wash Basins. Minor coal accumulations (Cherokee Coals) occur in this area above the thickest development of the basin sandy unit. This, again, suggests a direct relation between the position and thickness of coarse clastics and the location of peat swamps. The thick fluvial sandstone sequences result in favorable hydrology and a stable platform upon which thick peat could accumulate. Within the Vermillion and western Sand Wash Basin the upper shaly unit is a floodplain/paludal sequence with many fluvial channels (Roehler, 1965). Eastward, the paludal environment gives way to a more lowland fluvial environment that contains some sediments of a recurrent lacustrine environment (Roehler, 1965). A similar lithologic sequence may have been present in the eastern part of the Sand Wash Basin, however post-Paleocene uplift and erosion has removed much of this unit (Colson, 1969). Deposition of the Fort Union Formation ended in the early Eocene, when renewed tectonic activity caused the erosion of the margins of the Sand Wash Basin and the coarse sediments of the Wasatch Formation were deposited.

In summary, the distribution of Fort Union coals is controlled by the presence of thick fluvial channel-fill sandstones. Coal beds require hydrologic conditions necessary for initiation of peat swamps and a stable platform to promote organic accumulation. Fluvial channel sandstone belts were hydrologic and physiographic platforms for the accumulation of Fort Union coal beds. The thickest, most continuous coals occur in the lower coal-bearing unit in the central and eastern parts of the Sand Wash Basin, and they are potential coalbed methane targets. Coals of the upper shaly unit are not coalbed methane targets because of their thinness, lenticularity, and shallow burial depths.

Syntectonic Controls on Upper Cretaceous and Early Tertiary Sedimentation

The sedimentary tectonic model proposed herein for syntectonic and post-tectonic sedimentation in the Sand Wash Basin is based on the model of and text for syntectonic sedimentation for the Upper Cretaceous to early Miocene Cordilleran foreland described by Beck and others (1988). Characteristic syntectonic sedimentary facies within Cordilleran foreland basins include a narrow conglomerate facies adjacent to the thrust, a narrow sandstone/mudstone/coal facies just basinward, a basinal thrustward-thickening mudstone/coal facies above the depositional axis, and a wide distal sandstone/mudstone/coal facies (Beck and others, 1988). The wide distal sandstone/mudstone/coal facies depositionally thins above the shallowing basement opposite the impinging thrust or above the hanging wall of yet another thrust (Beck and others, 1988). Late Upper Cretaceous and early Tertiary sedimentation within the Sand Wash Basin is the result of similar syntectonic sedimentation and Laramide basement thrusting.

Thick-skinned basement thrusting along the Sierra Madre–Park Uplift caused the initial contemporaneous asymmetric uplift and subsidence at the Sand Wash Basin margins. During the *early thrusting phase* the uplift above the initially blind basement thrust along the Sierra Madre–Park Uplift was manifested as symmetric arching of Mesozoic and Paleozoic strata. Asymmetric basin subsidence due to thrust loading along the eastern margins of the Sand Wash Basin tilted the pre-Mesozoic strata toward the rising foreland uplift and caused erosional reworking of the sediments into thrustward (easterly) thickening wedges. Low-gradient fluvial and paludal depositional environments of the Fox Hills Sandstone and Lance Formation are characteristic of the basin's early phase of syntectonic deformation (figs. 49–51).

Continued early uplift associated with initial blind basement-thrusting of the Sierra Madre–Park Uplift also resulted in the deposition of the north-flowing fluvial massive K/T sandstone unit. Sandstone and coal occurrence maps of the massive K/T sandstone unit suggest that during the early phase of thrusting the Sand Wash Basin was characterized by both structural and depositional asymmetry. Both the structural and stratigraphic axes are located near the eastern margin (R93W) of the Sand Wash Basin. Subsequently, rapid thrusting terminated the deposition of the massive K/T sandstone and caused the development of an erosional unconformity.

During the *rapid thrusting phase,* the Sand Wash Basin was characterized by increased rates of thrusting, during which the rate of subsidence exceeded or equaled the rate of sedimentation. Characteristic syntectonic facies in the lower coal-bearing unit include a narrow, coarse conglomerate facies; clasts derived from Mesozoic strata, preferentially deposited adjacent to the impinging Sierra Madre–Park hanging wall thrust; a narrow sandstone/mudstone/coal facies immediately basinward; a thick sandstone/mudstone/coal facies above the structural axis; and finally a wide distal sandstone/mudstone/coal facies. Continued asymmetric subsidence during the rapid thrusting phase maintained a thrustward (easterly) dipping depositional gradient but caused the structural and depositional axes of the basin to migrate to the west in the Sand Wash Basin.

Recurrent rapid basement thrusting also caused the development of tectonically induced subsidence of the footwall of the Sierra Madre–Park thrust. Rapid subsidence adjacent to the frontal edge of the thrust relative to the rate of sedimentation is recorded by intraformational thickening of the gray-green mudstone unit (fig. 51), representing low-energy depositional environments, toward the impinging thrust. Localization of tributaries, flood plains and lacustrine environments (low-gradient, low energy) near the thrust, suggests that the frontal edges of the impinging hanging wall blocks did not supply the majority of sediment deposited into the Sand Wash Basin. Areas above the frontal ramps of basement thrusts however did provide coarse but volumetrically minor quantities of clastic (conglomerate) sediment to the basin.

Thrusting *slowed* or *ceased* during the transition phase between syntectonic and post-tectonic sedimentation, allowing the rate of sedimentation to equal or exceed the rate of thrust-load-induced subsidence adjacent to the frontal edge of the thrust. The gray-green mudstone, low-energy depositional environment was no longer localized near the frontal ramp of the basement thrust. Tectonically induced subsidence was no longer sufficient to maintain thrustward-dipping depositional gradients. Clastic sediment supply from the gently dipping westerly basin margins opposite the thrust decreased. The structural axes of the Sand Wash Basin migrated to the west, and the depositional axis became poorly defined during deposition of the gray-green mudstone unit. The low-energy floodplain, paludal and lacustrine depositional environments expanded to their greatest areal extent during this time. Slow or ceased thrusting was followed by a period of more rapid thrusting, resulting in the deposition of the basin sandy unit, with the new structural and depositional axes located near R96W.

Thrusting during the deposition of the basin sandy unit may have occurred along both the Uinta and Sierra Madre–Park Uplifts, resulting in the migration of the structural and depositional axes of the Sand Wash Basin to their present location near R96W. Sediment influx was from both the Uinta and Sierra Madre–Park Uplifts and the Sawatch Range. Subsequently, thrusting slowed or ceased and low-energy floodplain, paludal, and lacustrine depositional environments of the upper shaly unit expanded across the Sand Wash Basin. Low-gradient, low-energy conditions continued during the transition from syntectonic to post-tectonic sedimentation. The overlying Wasatch Formation represents the final phase of thrusting in the Sand Wash Basin, before the *post-thrusting phase* resulted in higher gradients, higher energy alluvial plains and fluvial environments of the Green River Formation that expanded across the Sand Wash Basin.

Conclusions

1. The major Tertiary coal-bearing unit and coalbed methane target in the Sand Wash Basin is the lower coal-bearing unit, Paleocene Fort Union Formation. Minor coal-bearing units and coalbed methane targets include the Upper Cretaceous Lance Formation and upper shaly units of the Paleocene Fort Union Formation.

2. The Lance Formation, the youngest Cretaceous stratigraphic unit in the Sand Wash Basin, overlies and intertongues with nearshore-marine deposits of the Fox Hills Sandstone and consists of nonmarine shales, lenticular sandstones, and coal beds. Coal beds are thicker and more abundant in the lower part of the Lance Formation above the Fox Hills Sandstone platform and range from a few inches (cm) to 10 ft (3.05 m). Locally, these coal beds have a limited lateral extent, can be traced for only a few hundred feet (m) in the subsurface, and may merge into single seams that are 15 to 20 ft (4.6 to 6.1 m) thick.

3. The Fort Union Formation is sand rich in the central and eastern portions of the basin.

4. The massive K/T sandstone, lower coal-bearing unit and basin sandy unit are present throughout the Sand Wash Basin and consist of continental-fluvial, lacustrine, and paludal deposits.

5. The lower coal-bearing unit of the Fort Union Formation contains north-trending fluvial sandstones and floodplain coal beds, which are laterally continuous. An increase in the suspended load carried by the fluvial system resulted in the building of levees that stabilized the channel axes and allowed for the formation of

extensive floodplains. Coal beds are thicker and more numerous in floodplain areas above and on the flanks of the thickest sandstone development. The lower coal-bearing unit contains some of the thickest individual coal beds (as much as 50 ft [15.2 m] thick) in the Sand Wash Basin. Net-coal thickness ranges from 0 to 80 ft (0 to 24.4 m) in as many as 12 seams at depths as much as 8,000 ft (2,438 m). Net-coal thickness and coal-seam continuity in the lower coal-bearing unit is greater than that in the upper shaly unit.

6. The gray-green mudstone unit is eroded in the center of the Sand Wash Basin, and the basin sandy unit lies disconformably on the lower coal-bearing unit.

7. The Cherokee coal zone, constituting the upper 250 to 450 ft (76.2 to 137 m) of the upper shaly zone occurs only in the west part of the Sand Wash Basin. The coal beds, about 3 to 10 ft (~0.91 to 3.05 m) thick, are not potential coalbed methane targets because of their thin and discontinuous nature and shallow burial depths. The Cherokee coal zone is removed by the Wasatch unconformity to the east.

8. Operationally defined lithostratigraphic correlations indicate that the Wasatch Formation contains no significant coals for coalbed methane exploration.

9. Sedimentation within the Upper Cretaceous to early Tertiary Sand Wash Basin is the result of syntectonic sedimentation and Laramide basement thrusting. Characteristic syntectonic sedimentary facies include a narrow conglomerate facies adjacent to the thrust, a narrow sandstone/mudstone/coal facies just basinward, a basinal thrustward-thickening mudstone facies, and finally a wide distal sandstone/mudstone/coal facies.

Coal Rank, Gas Content, and Composition and Origin of Coalbed Gases, Fort Union Formation, Sand Wash Basin

Andrew R. Scott

Abstract

Fort Union coal rank ranges from subbituminous along the basin margins to high-volatile A bituminous in deeper parts of the basin. Coal rank in T92–93W, where the thickest coals occur, is generally subbituminous to high-volatile C bituminous. Fort Union gas contents range from 9 to more than 300 ft^3/ton (0.3 to >9.4 m^3/t) but are generally less than 100 ft^3/ton (<3.1 m^3/t) (average 63 ft^3/ton [2.0 m^3/t]). The low gas contents of Fort Union coals are attributed to a combination of low coal rank and migration of gases in an active hydrodynamic system. Desorbed coalbed gases range from very wet to very dry (C_1/C_{1-5} values between 0.86 and 1.00) and have an average $C_1/_{1-5}$ value of 0.95. Fort Union coalbed gases are secondary biogenic, early thermogenic, or a mixture of these gas types.

Thermal Maturity and Gas Content

Both coalbed gas content and the composition of coalbed gases are influenced by the thermal maturity of the coals. Coal beds are unique in that the coal acts as both the source rock and the reservoir for hydrocarbons. Significant quantities of methane are generated from coals after the threshold of thermogenic methane generation is attained between vitrinite reflectance values of approximately 0.8 to 1.0 percent (Tang and others, 1991). Although the economic threshold of coalbed gas generation (~300 ft^3/ton [~9.4 m^3/t]) can be attained at vitrinite reflectance values of approximately 0.7 percent, higher vitrinite reflectance values are probably required to generate significant quantities of thermogenic methane. Gas contents for higher rank coals (high-volatile A bituminous and higher ranks) may exceed 400 to 500 ft^3/ton (>12.5 to 15.7 m^3/t) (Scott and Ambrose, 1992). Secondary biogenic and early thermogenic coalbed gases are associated with low-rank coals that have not reached the threshold of thermogenic gas generation (Scott, 1993). Primary biogenic gases, generated during peatification are probably not retained by the coal in significant quantities (Scott, 1993), whereas secondary biogenic gases are generated by bacteria that are introduced into the coals by meteoric waters flowing basinward from a recharge area. Gas contents associated with secondary biogenic methane generation are usually less than 100 ft^3/ton (<3.1 m^3/t). However, migration and conventional trapping of thermogenic and/or biogenic gases can result in unusually high gas contents in low-rank coals such as Fort Union coals in the Powder River Basin draped over channel sandstone belts (Law and others, 1991).

In **Kaiser, W. R., and others, 1993, Geologic and hydrologic controls on coalbed methane: Sand Wash Basin: The University of Texas at Austin, Bureau of Economic Geology, topical report prepared for the Gas Research Institute under contract no. 5091-214-2261 (GRI–92/0420), p. 107-113.**

Coal Rank

Proximate and ultimate data from four outcrop locations and 40 vitrinite reflectance values (Rm) from 15 Fort Union wells in the study area were used in developing a coal rank map on the base of the Fort Union Formation. All of the wells are located in Colorado along or east of the Little Snake River. Although vitrinite reflectance values ranged from 0.42 to 0.67, no distinct trend of increasing vitrinite reflectance with increasing depth was present (fig. 65). Vitrinite reflectance values remain constant or slightly decrease with increasing depth. This unusual vitrinite reflectance profile may be due to vitrinite reflectance suppression. Lower than expected vitrinite reflectance values have been attributed to suppression of reflectance when bitumen generated from hydrogen-rich macerals impregnates vitrinitic macerals during coalification (Hutton and Cook, 1980; Kalkreuth, 1982; Raymond and Murchison, 1990). Vitrinite reflectance suppression may begin at vitrinite reflectance values of 0.5 percent when the coals enter into the oil-generating stage. The amount of suppression will progressively increase with increasing maturation and/or increased amounts of hydrogen-rich macerals. Suppressed vitrinite reflectance values can range from 0.1 to 0.4 percent below expected or normal values (Levine, 1993). Fort Union coals are just entering the oil-generating stage and the constant or decreasing vitrinite reflectance values with increasing depth are probably due to vitrinite suppression. Assuming maceral composition does not change significantly, coals below 1,000 ft (305 m) above mean sea level have probably

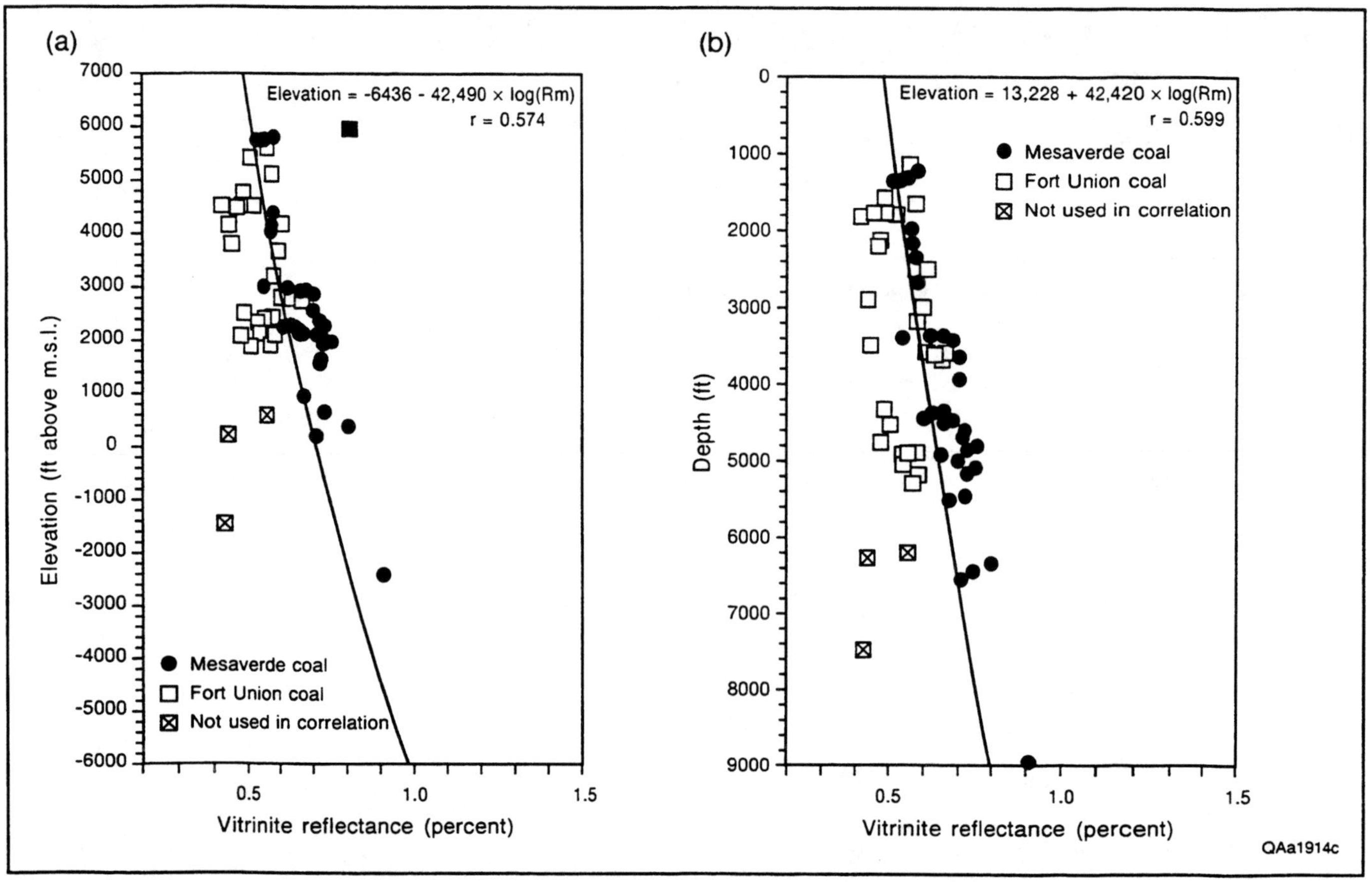

Figure 65. Fort Union vitrinite reflectance profiles relative to (a) elevation and (b) depth. Vitrinite reflectance values remain constant or slightly decrease with increasing depth. This unusual vitrinite reflectance profile may be due to vitrinite reflectance suppression from bitumen generated in the coals during coalification. Equations determined using both Fort Union and Mesaverde vitrinite reflectance data were used to construct a coal rank map of the base of the Fort Union.

generated more bitumen than coals above 2,000 ft (610 m) and, therefore, show more vitrinite suppression. Solidified bitumen is present in some samples below 2,000 ft (610 m) above mean sea level.

Suppression of vitrinite reflectance values prevented the use of vitrinite reflectance profiles of Fort Union coals to estimate coal rank in the western half of the basin. Therefore, the vitrinite reflectance curve calculated from Mesaverde coals in the Sand Wash Basin (Scott, this vol.) was used to estimate Fort Union coal rank in deeper parts of the basin. Coal rank ranges from subbituminous in outcrops along the southern, eastern, and northwestern margins of the basin to probably high-volatile A bituminous in the deeper parts of the Sand Wash Basin (fig. 66). Fort Union coals in the deeper parts of the Washakie Basin have reached the medium- to low-volatile bituminous ranks (Law, 1984). The lower coal rank along basin margins suggest that these structures formed or were forming during the main stage of coalification. Development of the Cherokee Arch also may have occurred before the main stage of coalification. However, due to the uncertaintly of vitrinite reflectance measurements in low-rank coals, the low rank of Fort Union coal beds in this area could make determination of the relationship between structural development and coalification difficult or impossible even if sufficient data were available.

Assuming that surface vitrinite reflectance values range between 0.2 and 0.3 percent (Teichmüller and Teichmüller, 1981), estimated vitrinite reflectance values of 0.35 to 0.40 percent in Fort Union outcrops indicate that uplift and erosion have removed some overburden from around the basin margins. The total amount of overburden could not be estimated from Fort Union vitrinite reflectance profiles (fig. 65). However, total overburden removal for the Fort Union along the basin margins was probably >1,500 to 2,000 ft (>457 to 610 m) (Tyler and others, this vol., fig. 49).

Gas Content

Gas content readings from 126 coal samples from 8 wells were used to evaluate the distribution of gas contents for the Fort Union Formation. All gas content readings were measured by the U.S. Bureau of Mines

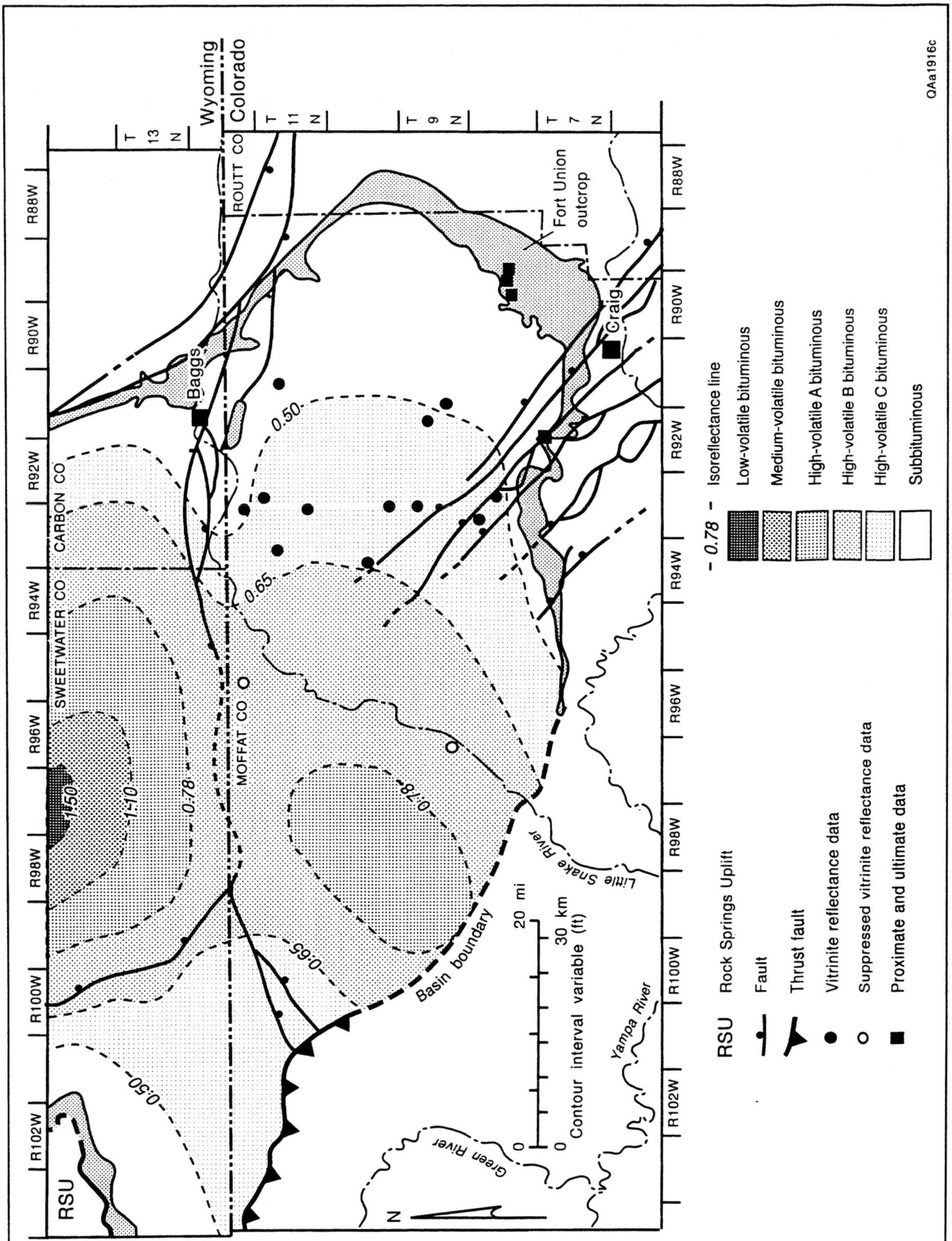

Figure 66. Coal rank map, base of the Fort Union. Net coal is thickest in T92–93, R9–13W.

method (Diamond and Levine, 1981) and were corrected to an ash-free basis when proximate data were available. In the absence of proximate data, all ash content values from the same well were averaged in order to correct the gas contents to a calculated ash-free basis. Ash content in the Fort Union ranges from less than 1 to 30.6 percent. Fort Union gas contents are generally less than 100 ft^3/ton (<3.1 m^3/t) (average 63 ft^3/ton [2.0 m^3/t]) and range from 9 to 301 ft^3/ton (0.3 to 9.4 m^3/t) (fig. 67).

Factors controlling gas content measurements include coal rank, basin hydrodynamics, localized pressure variations, sample type, sampling procedures, coal properties, analytical methods, and sample quality. Gas content profiles from other western basins generally show an increase in gas content with increasing coal rank, burial depth, and pressure. However, gas content profiles for Fort Union coal beds in the Sand Wash Basin are unusual because gas contents remain constant with increasing burial depth (fig. 68). Gas content does not change significantly among sample types (whole core, sidewall core, cuttings), suggesting that factors other than sample quality are affecting this profile. Vitrinite reflectance values also remain constant with increasing depth (fig. 65), suggesting that there is a relation between gas content and coal rank.

The unusual gas content profile may due to a combination of coal rank and basin hydrodynamics. Heads in the Fort Union Formation generally do not change significantly with stratigraphic interval or location across the basin, and the presence of relatively fresh water along basin margins suggests that large volumes of meteoric water and compactional fluids are moving though the Fort Union (Scott and Kaiser, this vol., figs. 76 and 77). The coals have not reached the thermal maturity level required to generate significant amounts of methane and minor amounts of thermogenic and/or secondary biogenic gases associated with the coals may have been flushed from the system by ground water. Low rank coals in the Powder River Basin where meteoric waters are flowing basinward from an eastern recharge area, have gas contents of less than 100 ft^3/ton (<3.1 m^3/t) (Scott and Ambrose, 1992) as do low-rank Mesaverde coals in the Sand Wash Basin.

Adsorption isotherms from the Timberlake Fee No. 1-20 well (T12N, R91W, Sec. 21, Colorado) and the Bridger coal mine (T12N, R100W) suggest that the maximum gas content possible for Fort Union coals is generally less than 300 to 400 ft^3/ton (<9.4 to 12.5 m^3/t) at reservoir pressures (fig. 69). Gas contents for coals in the Timberlake Fee No. 1-20 well range from 54 to 184 ft^3/ton (1.7 to 5.8 m^3/t) and average 99 ft^3/ton (3.10 m^3/t). Reservoir pressure is approximately 970 psi/ft (6,688 kPa/m) and the simple pressure gradient is 0.55 psi/ft (12.4 kPa/m), indicating that the Timberlake Fee No. 1-20 well is in an area of artesian overpressure. The adsorption isotherm and gas content data indicate that the coals are undersaturated with respect to methane. Assuming that gas contents for Fort Union coal beds are generally less than 100 ft^3/ton (<3.1 m^3/t) and the two isotherms shown in figure 68 are representative for the basin, then Fort Union coal beds throughout much of the Sand Wash Basin are probably undersaturated with respect to methane, and significant reduction of reservoir pressure will be required for methane desorption to occur. However, migration and conventional trapping of gases could result in higher-than-expected gas contents in some areas, and significant reduction of reservoir pressure may not be required to initiate gas production.

Most gas content measurements are performed at room temperature rather than at reservoir temperature. Since gas is desorbed more rapidly from coal surfaces at higher temperatures, gas contents measured at reservoir temperatures are usually higher than gas contents taken at room temperature. Gas contents for Fort Union coals determined at a reservoir temperature of 98°F (37°C) are approximately 1.2 times higher than gas contents made at room temperature (fig. 70). This value of 1.2 times is almost identical to the value estimated for Mesaverde coals at room and reservoir temperatures (Scott, this vol., fig. 33). Some gas contents taken at reservoir temperatures are significantly higher than expected and are to the left of the trend line (fig. 70).

Gas Composition

Coalbed gas composition is directly related to coal rank, basin hydrodynamics, and maceral composition (Scott and Kaiser, 1991; Scott 1993). The gas dryness index (the ratio of methane to methane through pentane; C_1/C_{1-5}) reflects the amount of chemically wet gases generated during the thermal maturation of hydrogen-rich coals. In general, when hydrogen-rich coals reach the oil-window or oil-generating stage (vitrinite reflectance of 0.5 to 1.2 percent) significant amounts of wet gases (ethane, propane, etc.) are generated, whereas coals having vitrinite reflectance values less than 0.5 percent or greater than 1.2 percent will generate relatively few wet gas components and have C_1/C_{1-5} values near unity (Scott and others, 1991a). The chemistry of coalbed gases can be significantly altered through biogenic activity. Bacterial alteration of chemically wet gases can remove nearly all of the wet gas components, producing chemically dry gases resembling thermogenic methane (James and Burns, 1984). Therefore, understanding basin hydrodynamics is important in evaluating coalbed gas origin. Furthermore, mixtures of biogenic and thermogenic

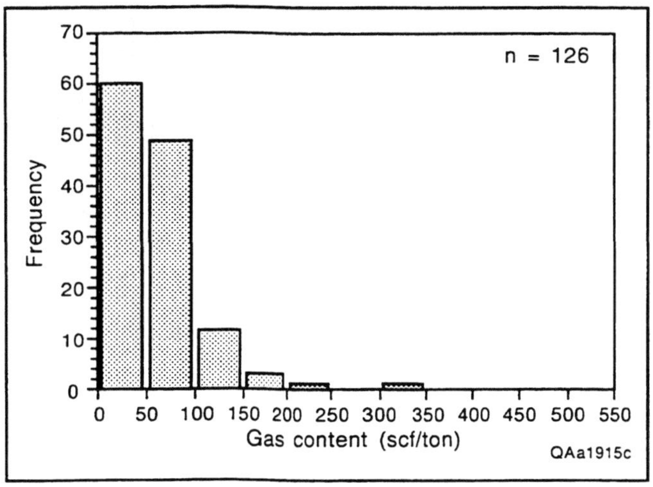

Figure 67. Histogram of gas content values, Fort Union coal samples.

coalbed gases are difficult to recognize using only gas dryness indices and methane isotopic data. The isotopic composition of carbon dioxide from coal beds may prove to be more useful in determining the biogenic or thermogenic nature of coal bed gases than methane isotopic data alone, particularly when mixtures of thermogenic and biogenic methane may be present.

The chemical composition of desorbed gas samples from 20 coal samples in 3 Fort Union wells were used to evaluate the chemical composition and origin of coalbed gases. Only three samples were analyzed for carbon dioxide. Although no produced coalbed gases were available for analysis in the basin, the compositional ranges of a large number of desorbed coalbed gases can be used to approximate the compositional ranges of produced gases (Scott, 1993). Desorbed gases often contain more carbon dioxide, nitrogen, and wet gas components (Mavor and others, 1991; Scott, 1993), particularly if higher (reservoir) temperatures are used during desorption. Although the higher temperatures used during desorption experiments desorb more carbon dioxide and wet gas components, these components remain strongly sorbed to the coal surface at reservoir pressures. Therefore, the produced coalbed gases are often chemically drier and contain less carbon dioxide than indicated by desorbed gas compositions.

Gas dryness indices for Fort Union coalbed gases range from 0.86 to 1.00 and average 0.95. These gas dryness indices are similar to gas dryness indices of desorbed coalbed gases in the Piceance Basin, which range from 0.79 to 1.00 and average 0.95 (Scott, 1993). Carbon dioxide content in Fort Union coal beds ranges from 4.2 to 6.9 percent and averages 5.4 percent. However, more gas compositional data are required to fully evaluate carbon dioxide content of Fort Union coals. Nitrogen content data were available for only three samples, and two of these sample contained more than 20 percent nitrogen indicating contamination with air. Nitrogen content for gases produced from subbituminous coals is probably variable. Nitrogen content in coalbed gases is generally highest in high-volatile C and B bituminous (Rm of 0.5 to 0.8 percent) and subsequently decreases with increasing coal rank (Scott, 1993). Therefore, produced Fort Union coalbed gases could have nitrogen contents ranging from 0 to more than 10 percent, depending on rank and maceral composition.

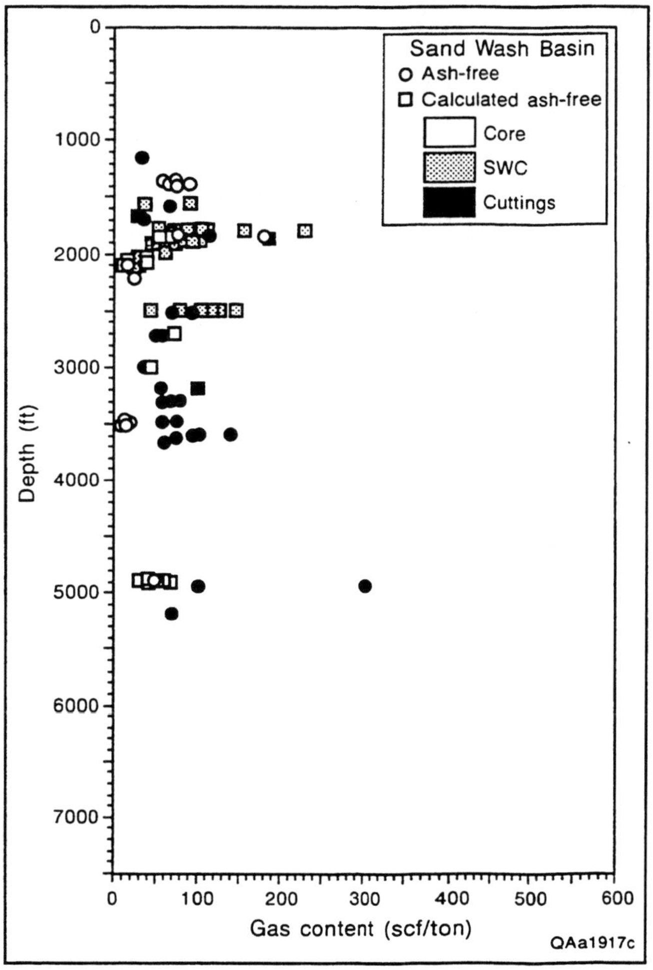

Figure 68. Gas content profile, Fort Union coals. The slightly higher gas content values at ~1,800 ft (~549 m) are from wells in T12N, R91W that correspond to an area of artesian overpressure.

Origin of Coalbed Gases

Early thermogenic, thermogenic, and secondary biogenic gases are associated with coal beds (Scott, 1993). Early thermogenic gases are generated during the early stages of coalification between vitrinite reflectance values of 0.5 and 0.8 percent, whereas significant quantities of thermogenic gases are not generated until the threshold of thermogenic gas generation is reached between 0.7

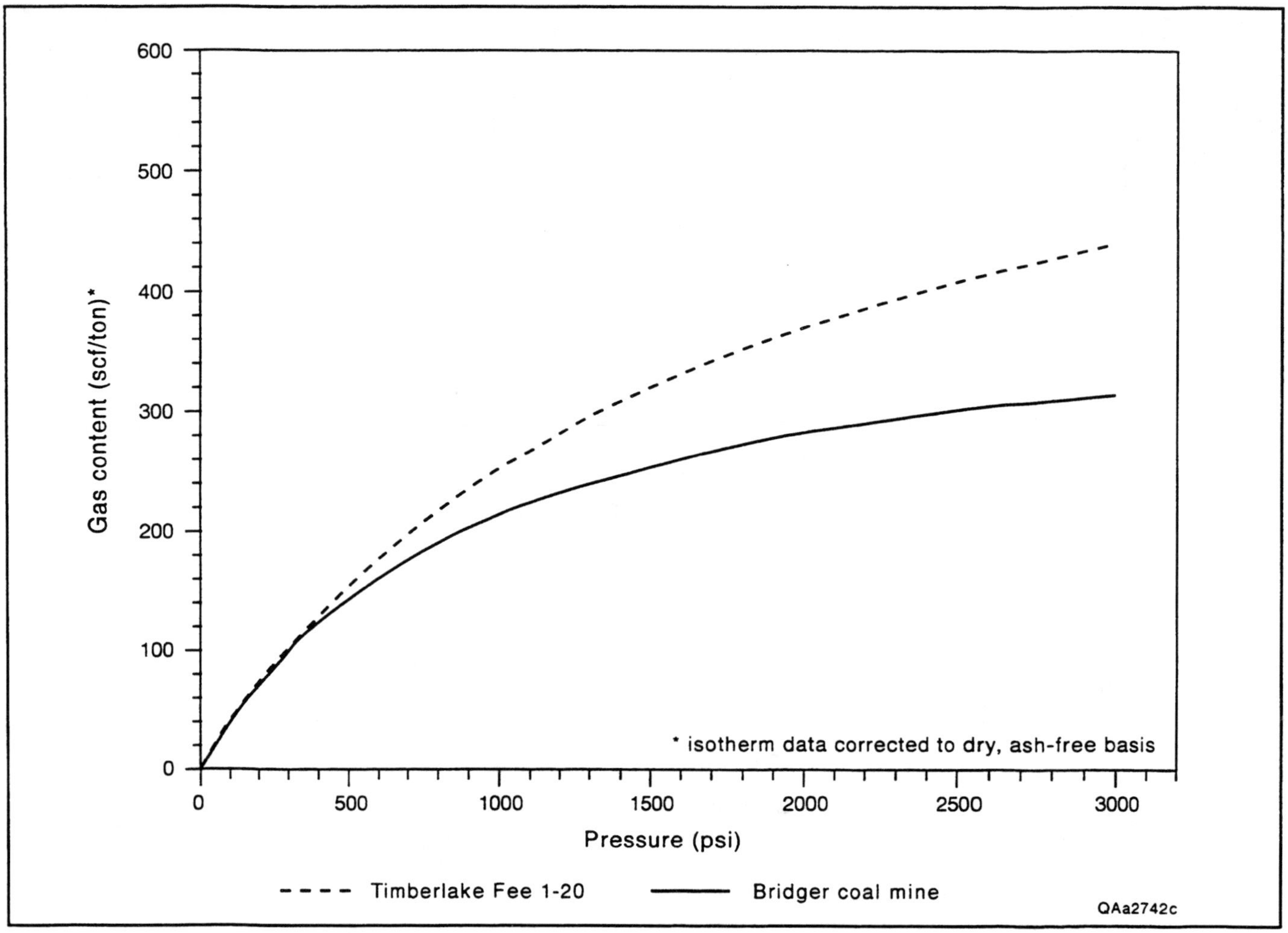

Figure 69. Adsorption isotherms for Fort Union coal samples. Isotherm number 1 is from the Timberlake Fee No. 1-20 well in the Sand Wash Basin. The second isotherm is from the Bridger coal mine along the Rock Springs Uplift (REI, 1990).

and 1.0 percent (Tang and others, 1991). Primary biogenic gases, generated during peatification, are probably not preserved in significant quantities (Levine, 1993; Scott, 1993). Secondary biogenic gases (Scott, 1993) are formed through bacterial degradation of chemically wet coalbed gases and organic compounds on the coal by bacteria transported by meteoric waters flowing basinward from a recharge area (Scott and others, 1991a, b; Kaiser and others, 1991b; Scott and Kaiser, 1991).

Determining the source of methane and carbon dioxide in coalbed gases is important for evaluating origin of coalbed gases and the migration of coalbed gases within the basin. Coal rank determines the quantities and types of thermogenic gases generated from a coal. Maximum thermogenic carbon dioxide generation occurs in lower rank coals, whereas maximum methane generation occurs in higher rank coals. Significant amounts of carbon dioxide are released from coals during maturation. Based on data from Levine (1992), more than 3,300 ft^3/ton (>103.4 m^3/t) of carbon dioxide are generated from a coal over the peat to high-volatile B bituminous ranks and an additional 600 ft^3/ton (18.8 m^3/t) is generated during the high-volatile A bituminous rank. Assuming that only carbon dioxide and methane are generated from a coal during coalification, approximately 680 ft^3/ton (21.3 m^3/t) of methane are generated between estimated vitrinite reflectance values of 0.25 and 0.60 percent (Levine, 1992). However, pyrolysis experiments on lignite performed by Tang and others (1991) suggest that coals generate less than 100 ft^3/ton (<3.1 m^3/t) over the same vitrinite reflectance range. Therefore, total methane production from high-volatile B bituminous and lower rank coals in the Fort Union Formation probably ranges between 50 and 500 ft^3/ton (1.6 to 15.6 m^3/t). Gas contents in Fort Union coals are generally less than 100 ft^3/ton (<3.1 m^3/t) (figs. 67 and 68), suggesting that a significant portion of the gas has migrated out of the system.

An additional source of methane and carbon di oxide and methane is from bacterial activity. Bacteria transported through permeable coals in ground water moving basinward generate secondary biogenic gases

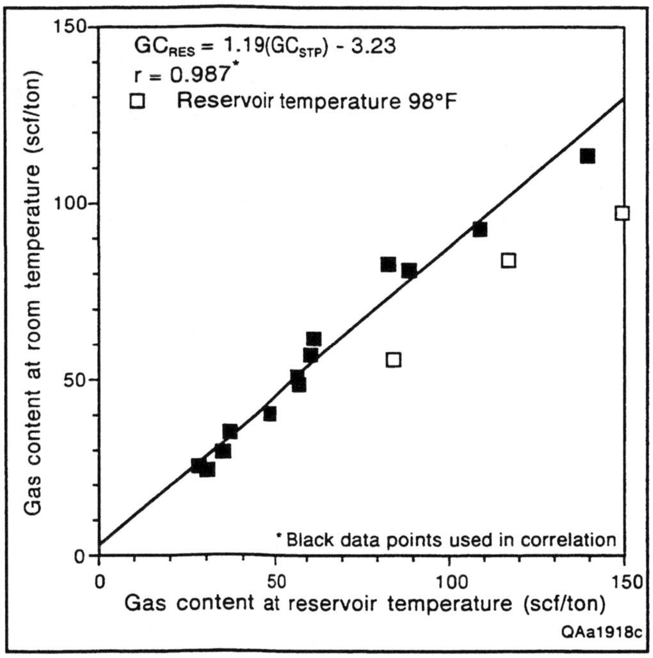

Figure 70. Gas content determined at room and reservoir temperatures.

(Scott, 1993). The origin of methane and carbon dioxide in coalbed gases can sometimes be determined from isotopic data. Methane in low rank coals with $\delta^{13}C$ less than –50 ‰ are probably secondary biogenic gases. However, coalbed gases with isotopic values between –40 to –55 ‰ can be biogenic, early thermogenic, or a mixture of both types. Thermogenic carbon dioxide released during coalification will be depleted in ^{13}C having $\delta^{13}C$ values of –25 to –15 ‰. Biogenic carbon dioxide is enriched in ^{13}C with $\delta^{13}C$ values ranging from –20 to +30 ‰ (Jenden, 1985) depending on the intensity and duration of bacterial activity. Therefore, carbon dioxide with positive $\delta^{13}C$ values is predominantly biogenic whereas $\delta^{13}C$ values less than –15 ‰ are generally thermogenic; mixtures of biogenic and thermogenic gases falling somewhere between. However, carbon dioxide derived from magmatic sources ($\delta^{13}C$ values of –7 to –9 ‰; Jenden, 1985) should not be ignored when evaluating gas origin.

Fort Union coalbed gases were not available for detailed isotopic analyses. However, coalbed gas origin can still be evaluated based on coal rank data and basin hydrodynamics. Although early thermogenic gases have probably been generated, vitrinite reflectance profiles suggest that Fort Union coals have not reached the thermal maturity level required to generate significant quantities of thermogenic gases. Meteoric water is flowing basinward (Scott and Kaiser, this vol.), suggesting that bacteria may have been introduced into the system and subsequently generated biogenic gases. Therefore, Fort Union coalbed gases are probably secondary biogenic and/or early thermogenic. Low coal rank, resulting in limited methane generation, and migration of biogenic and/or thermogenic gases out of the coals in an active hydrodynamic system probably explain the low gas contents in Fort Union coals. However, conventional and hydrodynamic trapping of migrating gases could result in higher than normal gas contents in some areas of the basin.

Conclusions

1. Fort Union coal rank ranges from subbituminous along the basin margins to high-volatile A bituminous in Sand Wash Basin and medium- to low-volatile bituminous in the Washakie Basin. Coal rank in the eastern part of the basin (T92–93N) where the thickest coals occur is generally subbituminous to high-volatile C bituminous; the coals are just entering the early stages of thermogenic gas generation.

2. Ash-free Fort Union gas contents range from 9 to more than 300 ft^3/ton (0.3 to >9.4 m^3/t) but are generally less than 100 ft^3/ton (<3.1 m^3/t) (average 63 ft^3/ton [2.0 m^3/t]). Gas contents made at reservoir temperatures (98°F [37°C]) are approximately 1.2 times higher than gas contents measurements taken at room temperature.

3. Factors controlling gas content distribution include coal rank, coal characteristics, localized pressure variations, basin hydrodynamics, and conventional trapping of migrating early thermogenic and biogenic gases. Low gas contents probably result from a combination of low coal rank and migration of gases in an active hydrodynamic system.

4. Fort Union coalbed gases desorbed from conventional cores range from very wet to very dry having C_1/C_{1-5} values ranging from 0.86 to 1.00; the average C_1/C_{1-5} value is 0.95. Fort Union coalbed gases are secondary biogenic and/or early thermogenic.

Hydrologic Setting of the Fort Union Formation, Sand Wash Basin

Andrew R. Scott and W. R. Kaiser

Abstract

The Fort Union Formation is part of an Upper Cretaceous/lower Tertiary regional aquifer system confined above and below by the Green River Formation and Lewis Shale, respectively. The distribution of artesian overpressure along the eastern part of the Cherokee Arch is controlled by faulting and facies distribution. Meteoric water probably enters the Fort Union outcrops in the Elkhead Mountains to the southeast of Baggs, Wyoming. Westward decrease in simple pressure gradient, salinity gradient, and higher heads in the coal-bearing unit suggest basinward movement of water. The potentiometric surfaces for the Fort Union and other hydrostratigraphic units in the Upper Cretaceous/lower Tertiary aquifer system do not change significantly across the Cherokee Arch, suggesting that lateral flow is sluggish. Vertical pressure gradients indicate a moderate to strong potential for upward flow along the Cherokee Arch. Gases generated from the Mesaverde Group, Lewis Shale, or Tertiary shales and coal beds in the deeper parts of the basin probably have migrated updip and vertically toward the Cherokee Arch and basin margins. In an interconnected regional aquifer system, dynamic equilibrium between updip and downdip fluid migration is attained and a flattened potentiometric surface results.

Introduction

The Fort Union Formation is part of a very thick, regional aquifer system that includes all strata between the Lewis Shale and Green River Formation (Tyler and Tremain, this vol., fig. 2); it is here termed the Upper Cretaceous/lower Tertiary aquifer system. Fort Union hydrology was evaluated in an analysis of hydraulic head, pressure regime, and hydrochemistry. Equivalent fresh-water heads were calculated from shut-in pressures (SIP) recorded in drill stem tests (DST). The pressure regime was evaluated on the basis of simple and vertical pressure gradients calculated from screened DST data. Chlorinity and total-dissolved-solid maps further defined ground water circulation patterns. Equivalent fresh-water heads were also determined for the Wasatch and Lance Formation and the Fox Hills Sandstone to better understand the larger Upper Cretaceous/lower Tertiary aquifer system. Fort Union hydrostratigraphy is reviewed as a prelude to a discussion of Fort Union hydrodynamics (hydraulic head, pressure regime, and hydrochemistry), which serves as the basis for interpretation of regional flow.

Hydrostratigraphy

Three hydrostratigraphic units in the Fort Union were evaluated: they are in descending order the upper shaly unit, basin sandy unit, and coal-bearing unit (Tyler and McMurry, this vol., fig. 48). The gray-green mudstone was not evaluated because it probably acts locally as an aquitard separating the basin sandy unit from the lower coal unit. The unamed massive K/T sandstone was included in the hydrologic investigation because there is no obvious aquitard separating the Fort Union from the massive sandstone and, therefore, it is probably in hydrologic communication with the overlying Fort Union. DST data and hydrochemistry for specific wells completed in the Wasatch and Lance Formations and the Fox Hills Sandstone were evaluated as needed. Heads for individual hydrostratigraphic units in the Upper Cretaceous/lower Tertiary aquifer system are similar and generally range from 6,000 to 7,000 ft (1,829 to 2,134 m) above mean sea level, suggesting that the system is a hydraulically interconnected aquifer system that behaves regionally as a single hydrologic unit.

Hydrodynamics

The hydrodynamics of the Fort Union Formation was established from equivalent fresh-water heads, formation fluid pressure, and hydrochemistry. Nearly 760 Fort Union DST data from 241 wells were taken from the Petroleum Information data base. DST data with simple pressure gradients less than 0.30 psi/ft (<6.8 kPa/m) were eliminated from the data base because of their uncertain validity, reflecting insufficient shut-in time, bad test data, presence of gas, pressure depletion, or a combination of these factors. Furthermore, a plot of elevation versus pressure suggested that data less than 0.30 psi/ft (<6.8 kPa/m) generally plots off of the main trend line (fig. 71). The quality of DST data was characterized as good if the final shut-in time was greater than 60 minutes; moderate, if the final shut-in time was 30–60 minutes; and unknown if the initial and/or final

In Kaiser, W. R., and others, 1993, Geologic and hydrologic controls on coalbed methane: Sand Wash Basin: The University of Texas at Austin, Bureau of Economic Geology, topical report prepared for the Gas Research Institute under contract no. 5091-214-2261 (GRI–92/0420), p. 115-125.

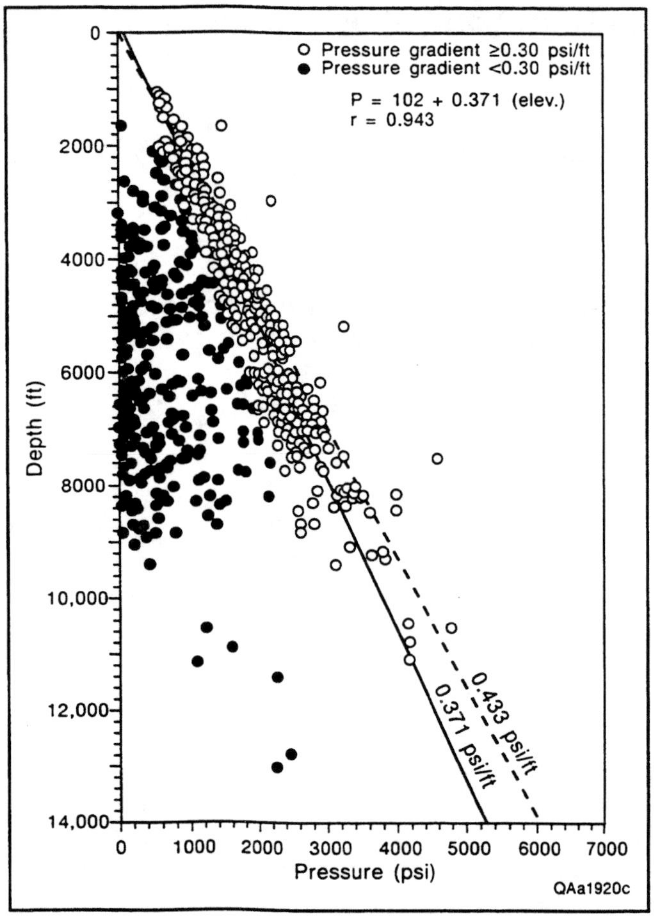

Figure 71. Pressure-depth plot, Fort Union DST data, Sand Wash Basin. DST data with simple pressure gradients less than 0.30 psi/ft (<6.8 kPa/m) were eliminated from the data base because of their uncertain validity.

shut-in times were not reported. Approximately 61 and 4 percent of the data were considered as good and moderate, respectively, whereas 35 percent were of unknown quality. Hydraulic heads and vertical pressure gradients were calculated from SIP's on a screened data set consisting of 450 DST's from 195 Fort Union wells. The stratigraphic test interval for 200 Fort Union DST's from 93 wells, and 36 massive K/T sandstone DST's for 15 wells was determined (fig. 72). Bottom-hole pressures were converted to pressure heads (BHP/hydrostatic gradient) using a fresh-water hydrostatic gradient of 0.433 psi/ft (9.8 kPa/m) and combined with elevation heads (kelly bushing minus midpoint of test) to obtain equivalent fresh-water heads.

Over 135 water analyses from 69 wells were available to evaluate basin hydrodynamics. The stratigraphy of 70 Fort Union samples intervals from 35 wells and 21 massive K/T sandstone samples from 11 wells were determined to evaluate water chemistry variability among the different stratigraphic units. Chemical analyses were dominantly of fluids recovered from DST's and secondarily of produced waters. The analyses were screened for analytical accuracy using, an ionic balance formula (Edmunds, 1981). In most cases, they balance exactly indicating that sodium and potassium were determined by analytical difference. Consequently, because of the nature of fluids analyzed and the exact ionic balance, the water analyses are of questionable validity and were used to delineate general concentration gradients rather than for detailed contouring of concentrations.

Potentiometric Surface

The potentiometric surfaces for the upper shaly unit, basin sandy unit, coal-bearing unit, and the massive K/T sandstone do not change significantly across the basin. Therefore, potentiometric-surface maps could not be confidently made for these units. A flattened potentiometric surface implies that lateral ground-ground water flow is sluggish. The potentiometric surfaces for the Wasatch and Lance Formations and the Fox Hills Sandstone also do not change significantly across the basin, indicating that lateral flow for the entire Upper Cretaceous/lower Tertiary aquifer system is also sluggish. Recharge occurs primarily along the eastern margin of the basin where annual precipitation over the Fort Union outcrop and subcrop exceeds 16 inches per year (41 cm/yr); annual precipitation along the northeastern and southern margins of the basin is less than 16 inches per year (12 to 16 inches/yr [20 to 41 cm/yr]) (Scott and Kaiser, this vol., fig. 41). Higher equivalent fresh-water heads in the coal-bearing unit along the eastern margin of the basin (T9-10N, R90W) probably reflect recharge from the elevated, wet eastern margin of the basin. Heads in the coal-bearing unit in T9-10N, R90W that exceed 7,000 ft (2,134 m) (fig. 73) reflect the topographic effect (elevation head) on total head. Recharge along the southern margin of the basin is limited by low annual precipitation and by local topography, which directs potential recharge southward towards the Yampa River. Burial of the Fort Union by thrust faults limits recharge from the southwest margin of the basin. Annual precipitation along the northwest margin of the basin is under 16 inches per year (41 cm/yr) (Scott and Kaiser, this vol., fig. 41), indicating that recharge from the southeastern edge of the Rock Springs Uplift is limited.

Pressure Regime

The average simple pressure gradient for the Upper Cretaceous/lower Tertiary aquifer system is 0.37 psi/ft (8.4 kPa/m) (fig. 71), indicating that the aquifer system is generally underpressured. Over 300 DST's from 4 study areas (fig. 72) were used to evaluate local variations

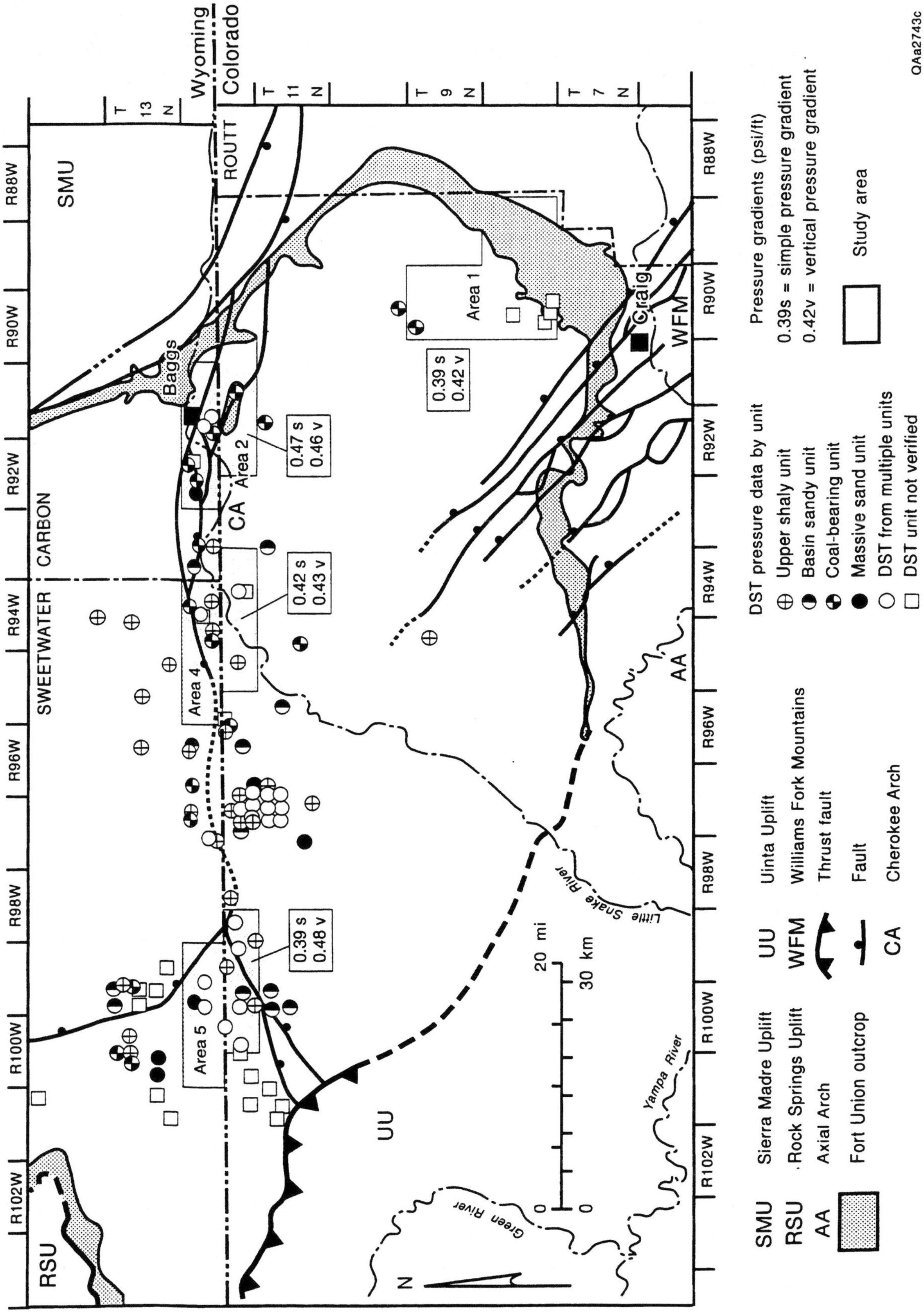

Figure 72. Fort Union pressure-analysis areas and DST well locations. The stratigraphic interval on more than 225 DST's in the Upper Cretaceous/lower Tertiary aquifer system was determined to evaluate variation in head and water chemistry among the hydrostratigraphic subunits of the aquifer.

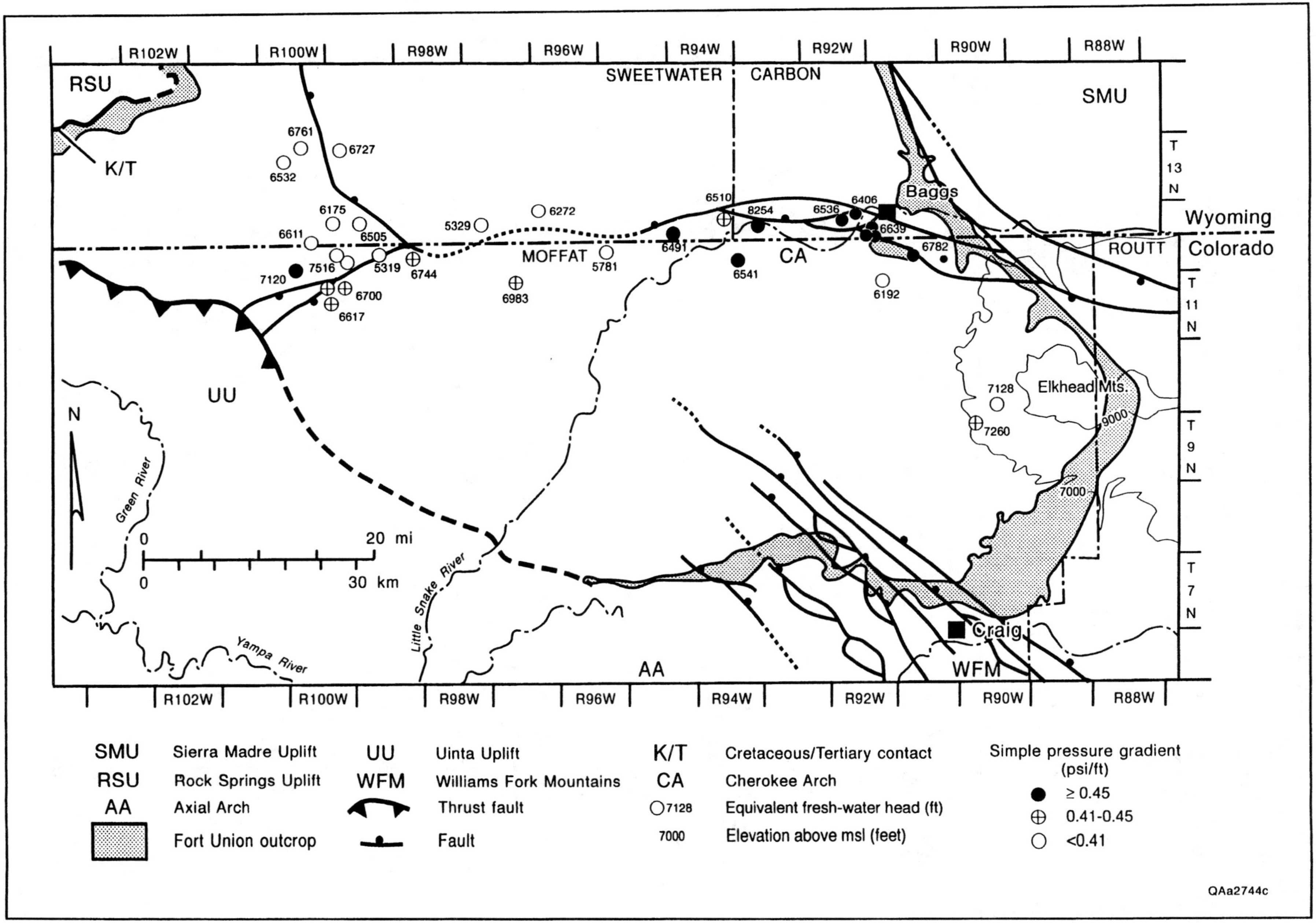

Figure 73. Equivalent fresh-water heads and pressure gradients in the lower coal-bearing unit, Fort Union Formation. The northwestward flow of water from the Elkhead Mountains and artesian overpressure are controlled by complex faulting along the Cherokee Arch and the distribution of coals and sandstones in the lower coal-bearing unit. Heads do not change significantly across the basin, suggesting sluggish lateral flow.

in pressure regime and to determine potential for vertical flow. To further evaluate the variability of upward flow potential across the basin, additional vertical pressure gradients were calculated for individual wells, which had three or more Fort Union DST's. Areas 1, 2, 4, and 5 contained a sufficient number of DST data to fully evaluate simple and vertical pressure gradients within the Upper Cretaceous/lower Tertiary aquifer system. Simple pressure gradients in Areas 1 and 5 are 0.39 psi/ft (8.8 kPa/m), whereas the simple pressure gradients in Areas 2 and 4 are 0.47 and 0.42 psi/ft (10.6 and 9.5 kPa/m), respectively. Underpressure in Area 5 probably reflects limited recharge off the Rock Springs Uplift, where annual precipitation is less than 10 to 12 inches per year (<25 to 30 cm/yr), whereas underpressure in Area 1 is less readily explained. Net sandstone thickness in the basin sandy unit and coal-bearing unit increases west of Area 1 (Tyler and McMurry, this vol., figs. 54 and 60), suggesting that underpressure may also reflect a draining effect of increased permeability or transmissivity downflow (Kaiser, 1993). Overpressure in Area 2 is probably artesian in origin with recharge occurring over the elevated, wet margins to the southeast. Simple pressure gradients for individual hydrostratigraphic units range from 0.45 psi/ft in the Wasatch to 0.49 psi/ft (10.2 to 11.1 kPa/m) in the Fort Union. The gray-green mudstone unit is thickest in the eastern part of of the basin (Tyler and McMurray, this vol., fig. 62) and may contribute to artesian overpressure by locally confining the lower coal-bearing interval. Although fracture flow may be promoted westward along the Cherokee Arch, faulting may act to compartmentalize the aquifer and actually impede westward flow, resulting in artesian overpressure. Close proximity to the recharge area results in higher simple pressure gradients in the Upper Cretaceous/lower Tertiary aquifer system than in the underlying Mesaverde (0.47 and 0.45 psi/ft [10.6 and 10.2 kPa/m], respectively). The simple pressure gradient in Area 4 is close to the hydrostatic gradient (0.433 psi/ft [9.8 kPa/m]), indicating that the Fort Union is normally pressured in this part of the basin.

The vertical pressure gradient, which is the slope of the pressure-elevation plot, is used to indicate vertical flow direction. Vertical pressure gradients were determined for the complete set of screened DST data for each study area and for individual wells within the study area (figs. 74 and 75). The vertical pressure gradients for Areas 1 and 4 are 0.42 and 0.43 psi/ft (9.5 and 9.7 kPa/m), respectively, indicating little potential for vertical flow in these areas. Analysis of individual wells indicates that vertical fluid movement may be more complicated than indicated in areal analysis. A vertical pressure gradient based on data from four wells adjacent to the outcrop in Area 1 indicates a strong potential for downward flow (0.29 psi/ft [6.6 kPa/m]), which is consistent with recharge, whereas the vertical pressure gradient for a well located in T9N, R90W, Sec. 9 suggests a strong potential for upward flow (0.46 psi/ft [10.4 kPa/m]). It is not known if this unusually high vertical pressure gradient is a local anomaly, a regional trend, or an artifact of the test data. Although the average vertical pressure gradient for Area 4 is equal to the hydrostatic gradient indicating horizontal flow, vertical pressure gradients for individual wells range from 0.44 to 0.61 psi/ft (10.0 to 13.8 kPa/m), indicating that some wells have a moderate to strong potential for upward flow.

The vertical pressure gradient of the Upper Cretaceous/lower Tertiary aquifer system in Area 2 (0.47 psi/ft [10.6 kPa/m]) indicates a strong potential for upward flow. Vertical pressure gradients for individual wells range from 0.33 to 0.46 psi/ft (7.5 to 10.4 kPa/m; fig. 75), depending on proximity to the outcrop and location within fault blocks. High vertical pressure gradients in Area 2 reflect aquifer confinement westward as faulting results in reservoir compartmentalization and resistance to lateral flow, which cause a buildup of reservoir pressure. The average vertical pressure gradient for the Upper Cretaceous/lower Tertiary aquifer system in Area 4 is 0.43 psi/ft (9.7 kPa/m) (fig. 74), indicating no potential for vertical flow. However, vertical pressure gradients for individual wells range from 0.44 to 0.66 psi/ft (10.0 to 14.9 kPa/m) with wells closer to the Cherokee Arch fault system having higher vertical pressure gradients and strong upward flow potential. This suggests that vertical fluid migration along the fault zone may be occurring. Higher pressures and local potentiometric mounds near the faults suggest the addition of fluids moving vertically upward along the faults.

The average vertical pressure gradient for all wells in Area 5 is 0.48 psi/ft (10.9 kPa/m), indicating a strong potential for upward flow. Vertical pressure gradients for individual wells ranged from 0.35 to 0.92 psi/ft (7.9 to 20.8 kPa/m), suggesting that fluid movement in this area is complex. Wells adjacent to faults on the east flank of a northeast-trending anticline have the highest vertical pressure gradients, whereas several wells on the anticline crest have lower vertical pressure gradients, suggesting the influence of free gas and crestal fracturing, which promotes pressure equilibration. High vertical pressure gradients suggest vertical migration of fluids up faults or updip to the anticline.

Hydrochemistry

Chlorinity, TDS, and calcium contents in the Upper Cretaceous/lower Tertiary aquifer system are variable

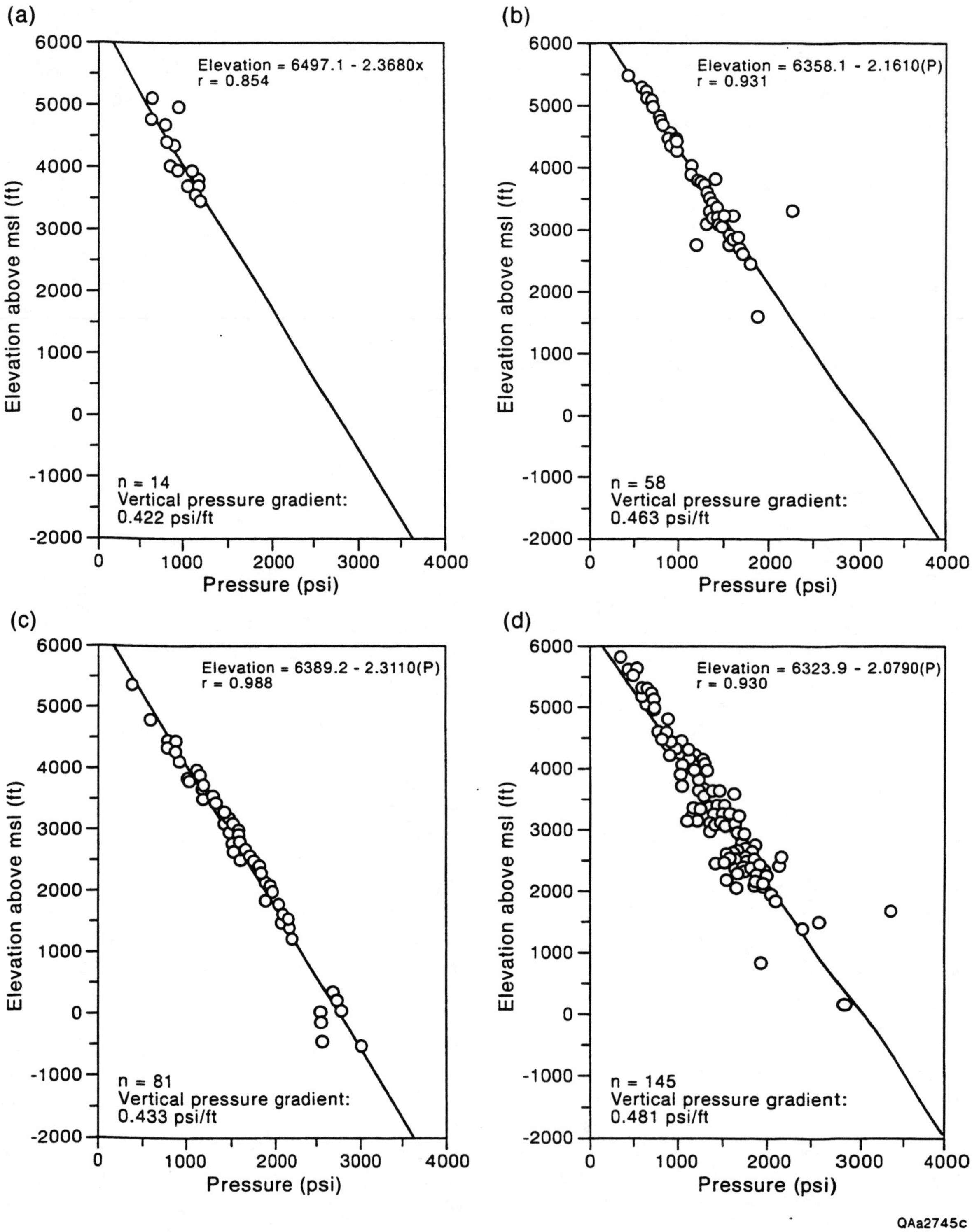

Figure 74. Fort Union pressure-elevation plots for four pressure-analysis areas. The large vertical pressure gradients in Areas 2 and 5 indicate a strong potential for upward flow, whereas Areas 1 and 4 indicate little potential for vertical flow.

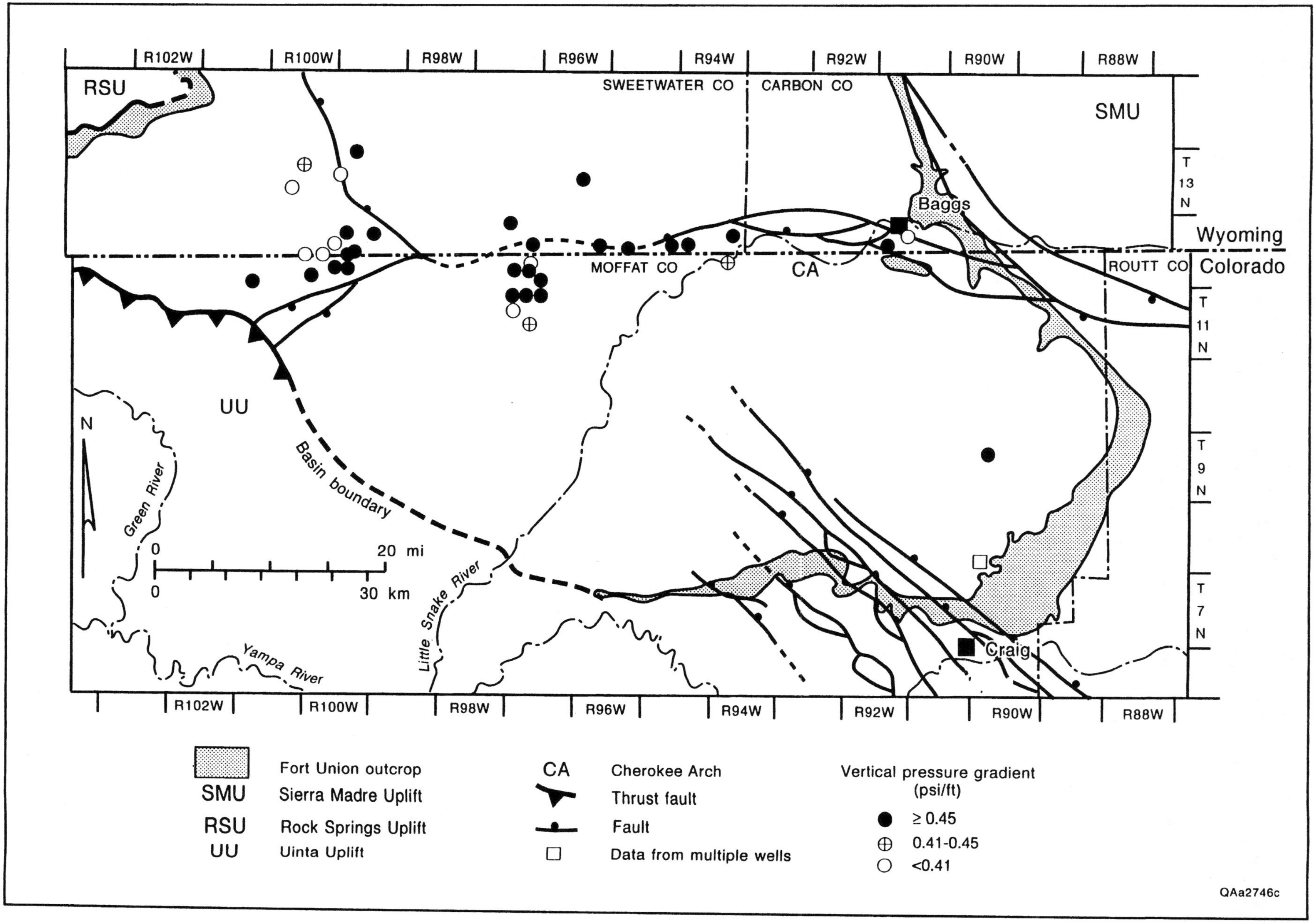

Figure 75. Distribution of vertical pressure gradients by well, Fort Union Formation. Most of wells located along the Cherokee Arch have vertical pressure gradients greater than 0.45 psi/ft (>10.2 kPa/m), suggesting a strong potential for upward flow. Compactional fluids and gases generated from the deeper parts of the basin may be converging on the arch.

across the basin. In order to evaluate changes in water chemistry across the basin and among individual hydrostratigraphic units in the aquifer system, the stratigraphic sampling interval of most of the water analyses was first verified from correlated geophysical logs. Water chemistry data were averaged for individual townships along the Cherokee Arch to determine the gross hydrochemical variation across the arch (fig. 76). Chlorinity and calcium contents are lowest along the eastern margin of the basin (generally less than 1,000 and 50 mg/L, respectively) but are generally higher in the deeper parts of the basin. High chloride and calcium contents are also associated with the Cherokee Arch fault system in R94W (fig. 76), suggesting that fluids have migrated vertically up the fault system. Hydrochemical data indicate that westward flow of water from the eastern margin of the basin is inhibited by the fault system in T94W. Vertical movement of fluids and mixing is indicated by the consistent average chloride contents in the Fort Union Formation adjacent to major fault systems. Chloride content is more variable among the different hydrostratigraphic units in the deeper parts of the basin, suggesting that mixing of formation waters among these units may not be limited.

Regional Flow

Ground water in the Sand Wash Basin flows mainly westward from the eastern recharge area in response to the topographic gradient and structural dip. The Elkhead Mountains (Scott and Kaiser, this vol., fig. 41) probably have a greater influence on ground water flow in the Fort Union than in the Mesaverde because the Fort Union crops out along the base of the mountain range. Ground water flowing north and northwest down topographic and structural gradient from the Elkhead Mountains, where annual precipitation exceeds 30 inches per year (75 cm/y) probably turns west-northwestward upon encountering the northwest-trending faults of the Cherokee Arch fault system (fig. 73). Orientation of net sandstone and coal trends in the coal-bearing unit also favor northwestward flow off of the mountains (Tyler and McMurray, this vol., figs. 54 and 61). The basinward extent of artesian overpressure, developed along the eastern part of the Cherokee Arch, is both promoted and controlled by the structurally complex Cherokee Arch fault system and depositional fabric. Although fracture flow may be promoted westward, faulting may serve to compartmentalize the aquifer system and impede westward flow and limit the extent of artesian overpressure. Wells with overpressure in the lower coal-bearing unit in R93-94W (fig. 73) correspond to areas of high net sand thickness, suggesting that sandstones affect the distribution of regional overpressure by focusing regional flow. Fracture flow northwestward from the outcrop along the Cedar Mountain fault zone is also possible although there are insufficient pressure and hydrochemical data to verify this possibility.

Lateral ground water flow in the Fort Union is sluggish, as shown by its flattened potentiometric surface (fig. 75), which may reflect a static or near static system, greater permeability and/or aquifer thickness downflow, horizontal flow directed upward or downward, cross-formational flow (leakage), or a system that has begun to generate hydrocarbons. Higher heads on the east near the Elkhead Mountains, vertical pressure gradients in excess of hydrostatic, and westward decrease of simple pressure gradients argue against a static system, whereas loss of coal beds westward (Tyler and McMurray, this vol., fig. 60) argues against increased permeability and/or aquifer thickness downflow. Therefore, the flattened potentiometric surface probably reflects leakage and hydrocarbon generation. Despite the fact that elevation head decreases basinward by over 5,000 ft (1,525 m) along the Cherokee Arch, total head changes very little, indicating considerable increase in pressure heads. In other words, loss of elevation head is compensated for by an increase in pressure head. This increase is thought to reflect considerable upward leakage of water and hydrocarbons from the deeper parts of the basin and along the Cherokee Arch fault system; fluids are being added to the system from below or generated from within.

Mesaverde heads basinward exceed those in the Fort Union by 2,000 to 3,000 ft (610 to 915 m), indicating that leakage from below is certainly possible but predicated on fracture flow across the Lewis Shale. There is a moderate to strong potential for upward flow in the Fort Union along the Cherokee Arch. Vertical pressure gradients in the Fort Union were determined for 35 wells (fig. 75) and over 70 percent of them had gradients greater than 0.46 psi/ft (>10.4 kPa/m). Actual vertical migration of fluids along faults is shown by consistent water chemistries (fig. 76) and higher heads along or adjacent to faults (figs. 73 and 77).

Vitrinite reflectance profiles suggest that coal beds and shales in the Fort Union Formation have reached the thermal maturity level to generate gas at elevations less than minus 2,000 ft (<–610 m) above mean sea level (Scott, this vol., fig. 66). Hydrocarbons generated from coal beds and shales in the Upper Cretaceous/lower Tertiary aquifer system and underlying units in the deeper parts of the basin have migrated updip and vertically toward the Cherokee Arch and basin margins. Fluids move up and out of the basin in response to pressure resulting from gas generation and compaction. Although Fort Union coals and shales in the deep Sand Wash Basin have reached the thermal maturity level required to generate early thermogenic gas, they have

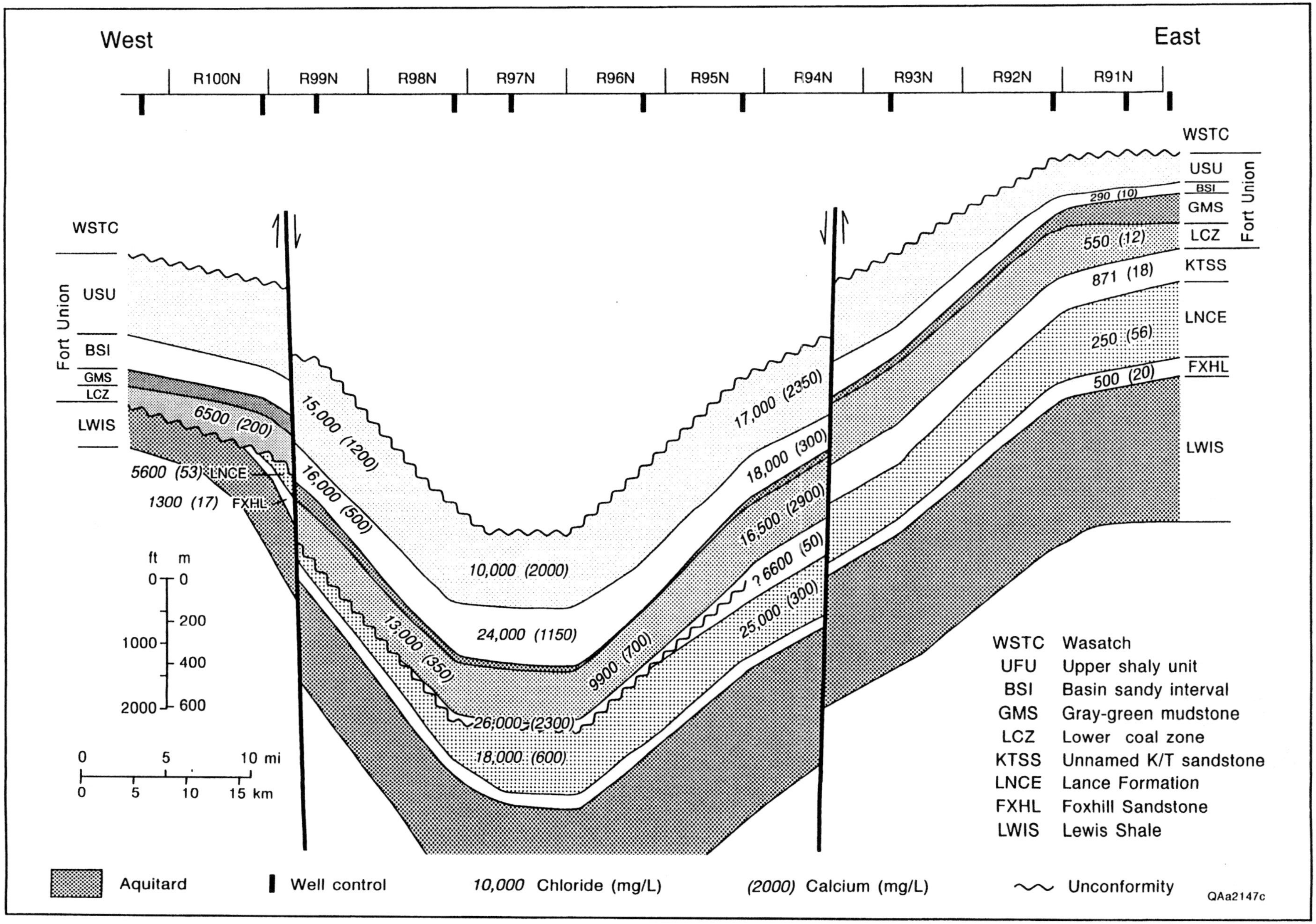

Figure 76. West-east hydrochemical cross section, Sand Wash Basin. The relatively consistent chloride content of hydrostratigraphic subunits in R94W suggests mixing of fluids migrating up faults. Relatively fresh waters are located near the outcrop in the eastern part of the basin. Line of section shown in Tyler and Tremain (this vol., fig. 1).

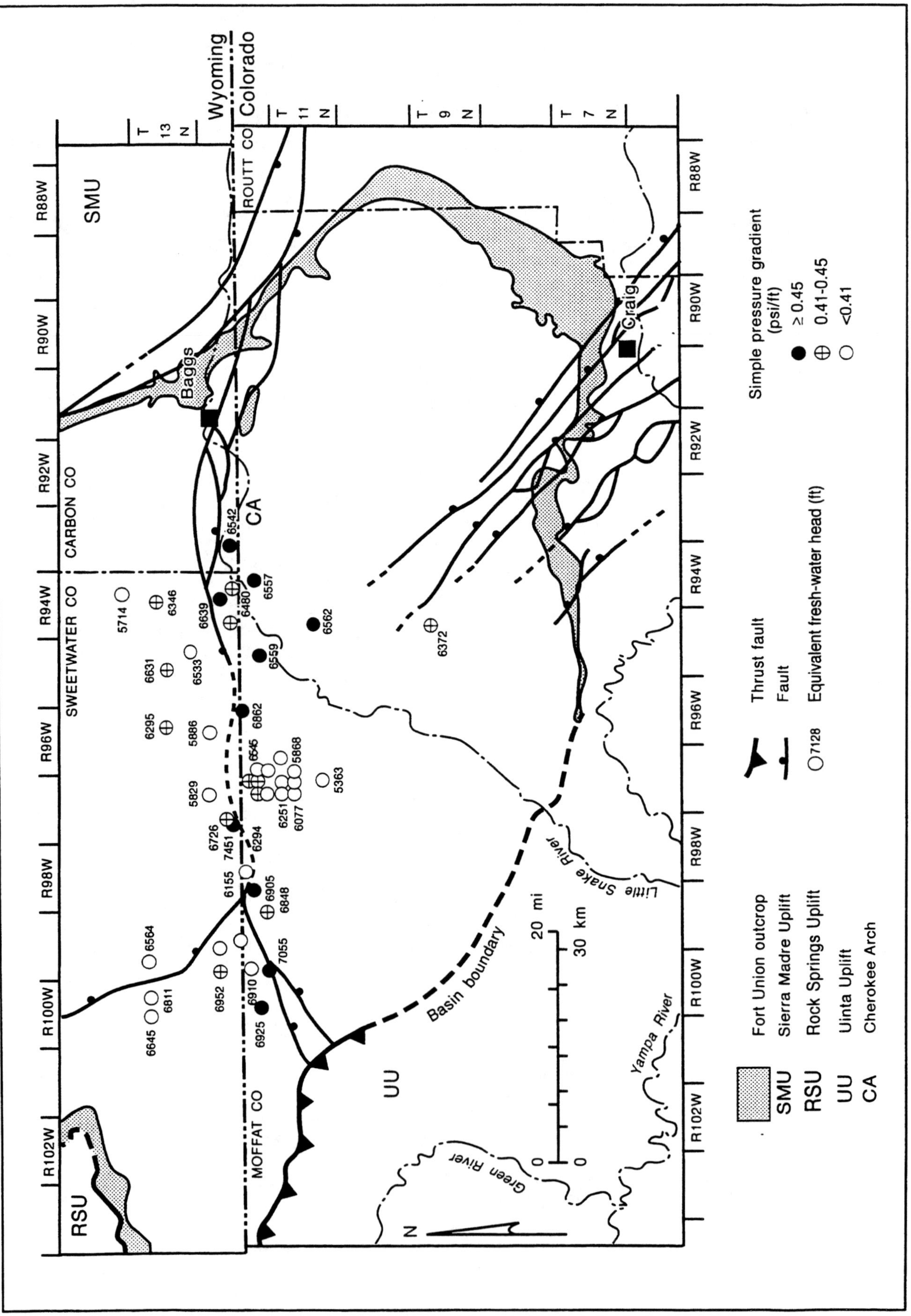

Figure 77. Equivalent fresh-water heads and pressure gradients in the upper shaly unit, Fort Union Formation. Wells with overpressure are generally located close to major fault trends, suggesting that localized overpressure may result from migration of gases and fluids upward along the faults. Heads do not change significantly across the basin, suggesting sluggish lateral flow.

probably not reached the thermal maturity level required to generate significant quantities of main-stage thermogenic gas. Temperatures in the Fort Union are generally less than 200°F (<93°C), suggesting that hydrocarbon overpressure is probably not present in the deeper parts of the basin. However, hydrocarbon overpressure may be present in the deeper parts of the Washakie Basin, where the Fort Union is more deeply buried and presumably has higher temperatures and lower permeablity conditions, which favor hydrocarbon overpressuring.

Conclusions

1. The Fort Union is part of the larger Upper Cretaceous/lower Tertiary regional aquifer system, which is confined below by the Lewis Shale and above by the Green River Formation. The potentiometric surfaces of individual hydrostratigraphic units within the regional aquifer system do not vary significantly among individual units or across the basin, indicating that lateral flow is sluggish. The larger regional flow system is near dynamic equilibrium among meteoric flow basinward, compactional flow up and out of the basin, upward leakage, and gas generation and migration. Consequently, the potentiometric surface along the Cherokee Arch, and possibly over a large part of the basin, is relatively flat.

2. Hydrochemical data, regional annual precipitation trends, head data, facies distribution, and fault geometry suggest that meteoric waters in the lower coal-bearing unit are derived from the Elkhead Mountains southeast of Baggs, Wyoming. Artesian overpressure in the Baggs area is controlled by faulting along the Cherokee Arch and facies changes in the lower coal-bearing unit. Water and gas are the pressuring fluids in the deeper parts of the basin.

3. Fluids in the Mesaverde Group and/or hydrocarbons generated from the Lewis Shale and/or coal beds and shales of the Upper Cretaceous/lower Tertiary aquifer system in the deeper parts of the basin have migrated updip and vertically toward the Cherokee Arch and basin margins. Vertical pressure gradients for selected study areas and individual wells indicate a moderate to strong potential for vertical flow along the Cherokee Arch and adjacent to major faults. Consistent water chemistry and higher heads along faults show that actual migration of fluids along faults has occurred.

Resources and Production

Resources and Producibility of Coalbed Methane in the Sand Wash Basin

W. R. Kaiser, Andrew R. Scott, Naijiang Zhou, Douglas S. Hamilton, and Roger Tyler

Abstract

Coalbed methane and coal resources in the Sand Wash Basin total 101 Tcf (2.86 Tm^3) and 377 billion short tons (342 billion t) and are 24 Tcf (679 Bm^3) and 180 billion tons (163 billion t) at shallow drilling depths of less than 6,000 ft (<1,830 m). Over 90 percent of the total gas resources and almost 75 percent of the total coal resources are in the Williams Fork Formation. Williams Fork coals yield large volumes of water and little gas, which is evident in a cumulative gas/water ratio of 15 ft^3/bbl (2.7 m^3/m^3). Similarly, Fort Union coals produced large volumes of water and essentially no gas. Production data were compared with geologic and hydrologic data to identify controls on coalbed methane production. Low average gas content (200 ft^3/ton [6.24 m^3/t]) and high water production (100's of bbl/d [10's of m^3/d]) are the major controls on production. Steep structural dip and coal distribution have restricted exploration to the eastern and southeastern margins of the basin. Prospective Williams Fork and Fort Union coals, respectively, lie basinward, coincident with convergent, upward flow along the leaky Cedar Mountain fault system, extending northwest of Craig, Colorado, and west along the Cherokee Arch into the Powder Wash field area. High productivity requires that geologic and hydrologic controls on production be synergistically combined. This synergism is evident in a comparison of the San Juan and Sand Wash Basins, where fundamental hydrogeologic differences between basins explain prolific and marginal production of coalbed methane, respectively, in the two basins. Out of that comparison, a basin-scale model for the producibility of coalbed methane is evolving. Its essential elements are ground-water flow through thick coals of high rank and high gas content orthogonally toward no-flow boundaries and conventional trapping of gas along them. The model remains to be tested and refined in other coal basins.

Introduction

Estimates of gas and coal resources rely on structure, topography, net-coal thickness, gas content, and ash content as reported earlier in this volume and published coal density data. These data were integrated to calculate gas and coal resources by geologic unit and drilling-depth fairway. This discussion of calculation methodology and resources is followed by a review of production, which has been mainly water and little or no gas. Geologic and hydrologic controls that contribute to marginal gas production are identified and summarized as a basis for suggesting prospective areas. Finally, in a comparison of the San Juan and Sand Wash Basins, we discuss the synergism required among controls for high productivity and propose a conceptual basin-scale model for coalbed methane producibility.

Resources

Gas and coal resources in the Sand Wash Basin were calculated using structure-contour and topographic maps, net coal thickness, gas content, coal density, and ash content and density. Net-coal thickness and area were combined to estimate net-coal volume, which was then used to calculate gas in place and coal tonnage, using gas content and ash-free coal density. Three resource estimates were made using: (1) no depth restrictions, (2) 7,500 ft (2,287 m), and (3) 6,000 ft (1,830 m). The basic equations used to calculate gas in place and coal tonnage are:

$$\text{GIP} = (h \times A) \times \rho_c \times GC \times C \qquad (1)$$

$$\text{TON} = (h \times A) \times \rho_b \times C \qquad (2)$$

where

GIP = gas in place (scf)
TON = coal tonnage (short tons)
GC = ash-free gas content (scf/ton)
h = coal thickness (ft)
A = area (square feet)
ρ_c = density pure coal (g/cc)
ρ_b = bulk density coal + ash (g/cc)
C = unit correction factor to convert to English units

In **Kaiser, W. R., and others, 1993, Geologic and hydrologic controls on coalbed methane: Sand Wash Basin: The University of Texas at Austin, Bureau of Economic Geology, topical report prepared for the Gas Research Institute under contract no. 5091-214-2261 (GRI–92/0420), p. 129-145.**

Based on considerations of pure coal density and weight and volume percentages of coal and ash content, these equations were modified for resource calculations. Coal resources were calculated using the coal bulk density (which includes both coal and ash), whereas in-place gas calculations should be made using pure coal density. Pure coal density is preferred for in-place gas resource estimates because the gas is assumed to be sorbed by the coal and not the ash. Failure to correct net-coal volume for ash content and using generalized bulk coal density data can result in an overestimation of in-place gas resources. Average ash contents for the Fort Union and Williams Fork Formations were 10.2 and 9.2 percent, respectively.

Previous studies have relied upon the coal density data from Averitt (1975), who only reports an average density of 1.32 g/cc (non-ash free) for all bituminous coals and 1.47 g/cc (non-ash free) for semianthracite and anthracite; data for medium- and low-volatile bituminous coals is not provided (Levine, 1993). Pure coal density ranges from 1.23 g/cc in bituminous coals to approximately 1.25 and 1.32 g/cc in low-volatile bituminous and semianthracite, respectively (Mahajan, 1989). Another potential source of error is confusion between volume percent and weight percentages of ash and coal. Coal and ash content values determined from proximate analyses are given in weight percent, indicating that coal thickness cannot be simply multiplied by weight percent coal to determine total coal volume. Net-coal volume must be multiplied by a volume correction factor that is based on the weight percentages of ash provided by proximate analyses. The total volume (V_t) of coal and ash (V_c and V_a, respectively) is given by $Vt = Vc + Va$, indicating that the volume fraction of coal (V_{fc}) and ash (V_{fa}) are determined by the following equations:

$$V_{fc} = \frac{Vc}{Vt} \tag{3a}$$

$$V_{fa} = \frac{Va}{Vt} \tag{3b}$$

The total density of the bulk coal (ρ_b), used in coal resource calculations, is derived by multiplying the volume fractions of coal and ash by the densities of pure coal and pure ash, respectively:

$$\rho_b = (V_{fc} \times \rho_c) + (V_{fa} \times \rho_a) \tag{4}$$

Pure coal density was related to a depth plot of pure coal density versus percent carbon in Levine (1993), which was first converted to equivalent vitrinite reflectance values, and then correlated with depth using the vitrinite reflectance profile equation in figure 28a to attain a pure coal density versus depth (D) equation:

$$\rho_c = 1.219 + 8.31 \times 10^{-6}(D) - 1.73 \times 10^{-9}(D^2) + 9.98 \times 10^{-14}(D3) - 4.06 \times 10^{-19}(D^4) \tag{5}$$

The density of pure coal is significantly less than the density of ash found in coal beds, indicating that the volume fraction of ash in a coal bed is often significantly less than weight percent of ash. For example, a high-volatile A bituminous coal containing 20 percent by weight ash will have coal (w_c) and ash (w_a) weight fractions of 0.80 and 0.20, respectively, whereas the coal and ash volume fractions will be 0.90 and 0.10, respectively. Therefore, coal volume must be corrected using volume correction factors rather than weight correction factors. The volume fraction of coal can be determined from the weight fractions of coal and ash using:

$$V_{fc} = \frac{1}{1 + \left(\frac{\omega_a \times \rho_c}{\omega_c \times \rho_a} \right)} \tag{6}$$

Bulk coal density values (r_b) are used in determining coal resources (equation 2). Bulk coal density values determined by Averitt (1975) represent generalized density values for two major coal ranks, suggesting that a direct relationship between pure coal density and bulk coal density based on experimental data would be more appropriate in determining coal resources over various ranks. Bulk coal density can be determined from pure coal density, ash density, and weight fractions of coal and ash using the following equation:

$$\rho_b = \rho_a + \frac{(\rho_c - \rho_a)(\rho_a \times \omega_c)}{(\rho_a \times \omega_c) + (\omega_a \times \rho_c)} \tag{7}$$

Regional changes of bulk coal density with increasing depth are determined by substituting values derived from equation 5 into equation 7. Regional variations of ash content can also be handled in a similar way if sufficient proximate analyses are available.

Coal and in-place gas resource calculations must include the appropriate treatment of bulk coal and ash density, pure coal density, and the weight and volume fractions of coal and ash. Resources can be overestimated if these factors are not considered. Modified gas-in-place and coal resource equations were derived by combining equations 5, 6, and 7 with equations 1 and 2 to yield:

$$\mathrm{GIP} = (h \times A \times V_{cf}) \times \rho_c \times GC \times C \qquad (8)$$

$$\mathrm{TON} = (h \times A) \times \rho_b \times C \qquad (9)$$

Structure maps on the top of the massive K/T sandstone (base of the Fort Union) and top of the Williams Fork Formation (Tyler and Tremain, this vol., figs. 5 and 6) and topographic maps were digitized and converted to a grid on Radian CPS software that utilized an evenly spaced node system with 9,840 ft (3,000 m) between nodes. This node spacing was selected because it was the smallest grid size (3.5 mi^2 [9.1 km^2]) that accurately reflected structure and mapped coal thickness, while at the same time minimizing computer calculation time. Drilling-depth fairways to 7,500 ft (2,287 m) and 6,000 ft (1,830 m) were defined by subtracting Williams Fork and Fort Union structural elevations from surface elevations. Gas and coal resource estimates were made for the lower coal-bearing unit of the Fort Union Formation and and Units 1 through 4 of the Williams Fork Formation, the basin's major coal-bearing stratigraphic units. Depths were calculated to the midpoint of each of those units, assuming average thicknesses for each. Depth to the midpoint of the lower coal-bearing unit of the Fort Union Formation was estimated by adding 300 ft to the structure map on the top of the massive K/T sandstone unit (Tyler and McMurry, this vol., fig. 60) and then subtracting this midpoint elevation from approximate ground level at each grid node. The elevation midpoint for Williams Fork Unit 4 was estimated by subtracting 200 ft (61 m) from the structure map on the top of the Williams Fork Formation (Tyler and Tremain, this vol., fig. 5), whereas the elevation midpoints in Units 3 through 1 (descending order) were estimated by subtracting an additional 400 ft (122 m) for each subsequent unit. Net coal within each unit is assumed to occur as an aggregate thickness at the midpoint of the unit.

Although maximum gas content increases with depth (fig. 78), reflecting higher reservoir pressures, values are erratic and may range from 50 to 550 ft^3/ton (1.56 to 17.16 m^3/t) over a small interval and vary significantly between coals separated by 10 ft (3 m) or less in the same well (Scott, this vol.). Because the amount of gas generated from a coal is a function of its burial history, maximum temperatures experienced, and maceral type, present-day burial depths do not always correlate with the amount of gas sorbed on the coal surface. For example, the Elkhead Mountains in the eastern part of the basin represent a topographic high, but there is no evidence to indicate that coal-rank (and gas content) in this area is significantly higher. Therefore, a gas content versus depth plot may predict erroneously high gas contents for this area even though the coals have probably not reached the thermal maturity required to generate significant amounts of gas. Consequently, we initially plotted gas content versus elevation. However, this approach led to an overestimation of gas content in shallower units and forced us to use the traditional plot of gas content versus depth. Our initial plot was made using gas-content values from whole core data from the combined Fort Union Formation and Mesaverde Group averaged over successive 1,000-ft (305-m) intervals. The Fort Union data were included because there were too few core-derived values to establish a separate Fort Union curve. However, this approach led to an overestimation of Fort Union gas contents. Therefore, separate plots of gas content versus depth were made, using a moving average over 1,000-ft (305-m) intervals with 500-ft (164-m) overlap (fig. 78). The Fort Union plot includes all gas-content data because Fort Union sidewall core and cuttings do not show distinctly lower gas-content values than those from core, as does the Mesaverde data. The increase of gas content with depth (D) used for in-place gas calculations for the Fort Union and Williams Fork Formations, respectively, are given by:

$$GC = 0.0120(D) + 16.6 \text{ (Fort Union)} \qquad (10)$$

$$GC = 0.0612(D) - 15.3 \text{ (Williams Fork)} \qquad (11)$$

Resources were calculated from equations 8 and 9. Gas contents (equations 10 and 11) were assigned to each grid node and corresponding coal volume by stratigraphic unit to calculate total resources for the basin and resources at depths of less than 6,000 ft (<1,830 m) and 7,500 ft (<2,287 m) (tables 4–7). Coalbed methane and coal resources in the Sand Wash Basin total 101 Tcf (2.86 Tm^3) and 377 billion short tons (342 billion t) and are 24 Tcf (680 Bm^3) and 180 billion tons (163 billion t) at shallow drilling depths of less than 6,000 ft (<1,830 m). Despite low average gas contents in the basin (Scott, this vol.), total gas resources are large because coal resources are large (tables 4 and 5). Reduction in gas resources of 56 to 76 percent occur when drilling depth restrictions are applied and are a consequence of steep structural dip. The Williams Fork is the most important gas- and coal-bearing unit evaluated; it contains 94 Tcf (2.66 Tm^3) and 280 billion tons (254 billion t) of coal, accounting for 93 and 74 percent, respectively, of the basin's total resources. Among stratigraphic units in the Williams Fork, Unit 4 has the most total gas (table 6), reflecting deep coal of high gas content. However, at depths of less than 6,000 ft (<1,830 m), Unit 1 has the most gas and is the richest coal-bearing unit (table 7). The Fort Union contains 7 and 26 percent of the total gas and coal

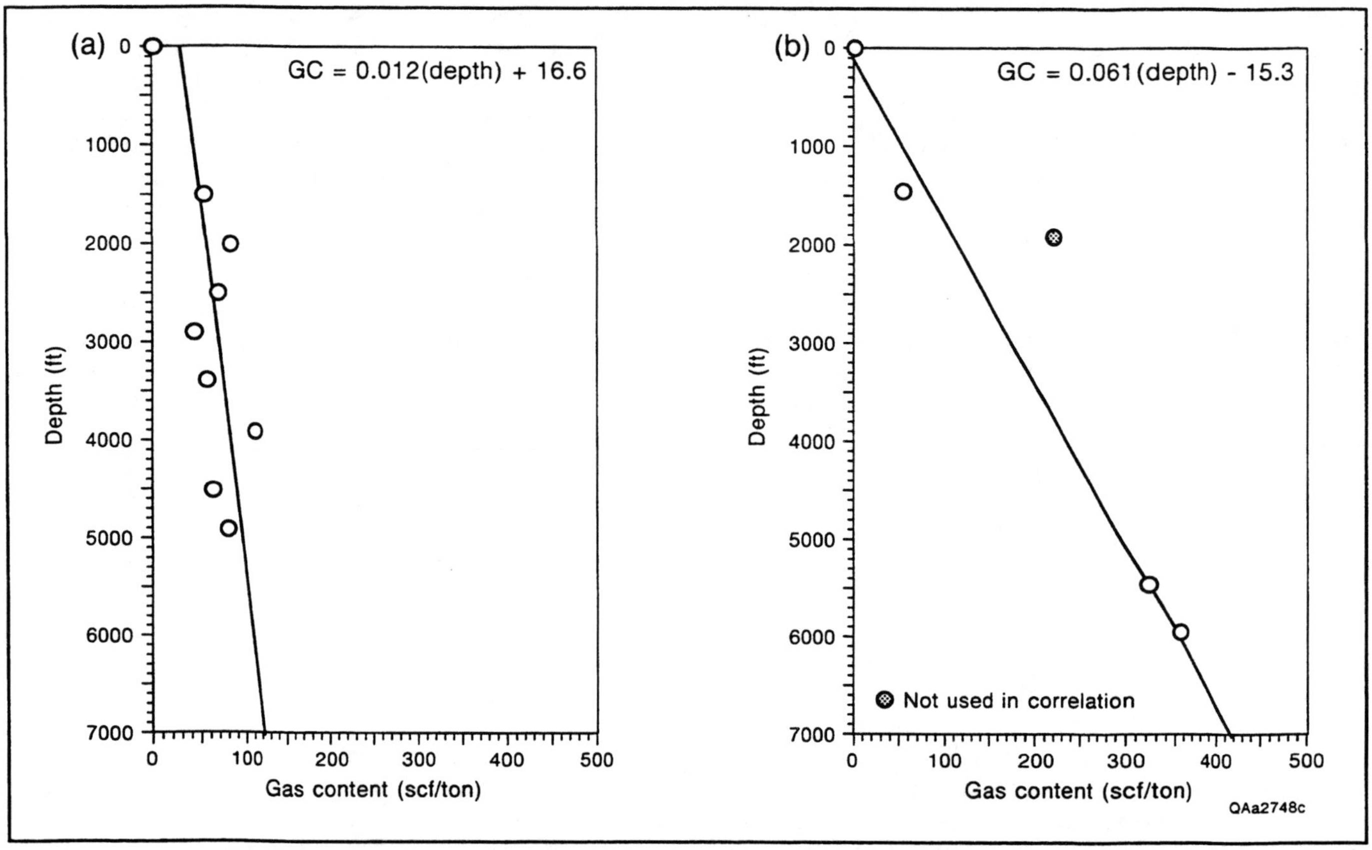

Figure 78. Gas content profiles and equations used for in-place gas and coal resource calculations. A moving average of 1,000 ft (305 m) intervals with 500 ft (153 m) overlap was used to determine a general relation between gas content and depth for the (a) Fort Union and (b) Williams Fork Formations.

resources and is least affected by drilling depth restrictions.

Production

Analysis of Williams Fork and Fort Union production is based on Petroleum Information reports (Petroleum Information, 1993), Dwight's Oil and Gas drilling histories, Colorado Oil and Gas Conservation Commission well completion updates, and operator records. Gas production from three Williams Fork fields has been minimal, whereas water production has been excessive (table 8). Cumulative gas and water production through December 1992 were 84 MM ft^3 (2.38 MM m^3) and 5.5 million barrels (0.87 million m^3) for a cumulative gas/water ratio of approximately 15 ft^3/bbl (~2.7 m^3/m^3). Only the Dixon field has produced gas for a cumulative gas/water ratio of approximately 25 ft^3/bbl (~4.4 m^3/m^3). There are 11 wells in Dixon field (fig. 79); 3 structurally high wells currently produce gas at rates of less than 50 Mcf/d (<1.4 Mm^3/d). Initially, eight wells were flowing artesian and served as dewatering wells; they flowed at rates ranging from 600 to 1,000 bbl/d (95 to 159 m^3/d) for a per-well average of approximately 700 bbl/d (~111 m^3/d) in 1991. Upon subsequent production, rates have declined to approximately 400 bbl/d (~64 m^3/d).

There are 16 plugged and abandoned wells in Craig Dome field (fig. 79). The wells were abandoned because the Williams Fork coals had low gas contents and could not be economically depressured (dewatered). They were produced 12 to 18 months with minor pressure drawdown and never produced gas. In 1991, water production per well ranged from 200 to 1,000 bbl/d (32 to 159 m^3/d), and averaged about 500 bbl/d (80 m^3/d); two were flowing artesian wells. The one Williams Fork well in Lay Creek field tested initially for 74 Mcf/d (2.1 Mm^3/d) and 800 bwpd (127 m^3/d). During production testing it produced 80 to 100 Mcf/d (2.3 to 2.8 Mm^3/d) and 100's of bwpd (10's of m^3/d). The Van Dorn well (T7N, R90W, sec. 29) made 100 Mcf/d (2.8 Mm^3/d) upon swabbing after an unsuccessful frac job and then died.

In 1989 and 1990, nine Fort Union coalbed wells were completed, production tested, plugged, and abandoned. During test periods ranging from 9 days to 7 months, the wells made zero to negligible volumes of gas and 10,000's of bbl of water (1,000's of m^3); one

Table 4. Coalbed methane resources of the Sand Wash Basin.

	Gas in Place (Tcf)		
	No. depth restriction	≤7,500 ft[a]	≤6,000 ft[a]
Fort Union[b]	7	4	3
Williams Fork[c]	94	40	21
Total:	101	44	24
Colorado			
Fort Union[b]	4	3	2
Williams Fork[c]	75	35	18
Total for Colorado	79	38	20
Wyoming			
Fort Union[b]	3	1	1
Williams Fork[c]	19	5	3
Total for Wyoming	22	6	4

Table 5. Coal resources of the Sand Wash Basin.

	Gross tonnage in billions of short tons		
	No. depth restriction	≤7,500 ft[a]	≤6,000 ft[a]
Fort Union	97	72	58
Williams Fork	280	175	122
Total:	377	247	180
Colorado			
Fort Union	61	50	42
Williams Fork	230	150	103
Total for Colorado	291	200	145
Wyoming			
Fort Union	36	22	16
Williams Fork	50	25	19
Total for Wyoming	86	47	35

[a]depth to base of coal-bearing unit
[b]based on average ash content of 10.9 percent
[c]based on average ash content of 9.2 percent

Table 6. Coalbed methane resources in the Williams Fork Formation. Units are listed in stratigraphic order.

	Gas in Place (Tcf)[a]		
	No. depth restriction	≤7,500 ft[b]	≤6,000 ft[b]
Unit 4	31	10	5
Unit 3	19	5	3
Unit 2	16	7	2
Unit 1	28	18	11
Total:	94	40	21
Colorado			
Unit 4	22	8	4
Unit 3	13	4	2
Unit 2	13	6	2
Unit 1	27	17	10
Total:	75	35	18
Wyoming			
Unit 4	9	2	1
Unit 3	6	1	<1
Unit 2	3	1	tr
Unit 1	1	1	1
Total:	19	5	3

Table 7. Coal resources in the Williams Fork Formation. Units listed in stratigraphic order

	Gross tonnage in billions of short tons		
	No. depth restriction	≤7,500 ft[b]	≤6,000 ft[b]
Unit 4	78	39	25
Unit 3	49	24	16
Unit 2	45	25	15
Unit 1	108	87	66
Total:	280	175	122
Colorado			
Unit 4	59	32	21
Unit 3	34	17	11
Unit 2	38	22	13
Unit 1	99	79	58
Total:	230	150	103
Wyoming			
Unit 4	19	7	4
Unit 3	15	7	5
Unit 2	7	3	2
Unit 1	9	8	8
Total:	50	25	19

[a]based on average ash content of 9.2 percent; [b]depth to base of coal-bearing unit; tr = less than 0.5 Tcf

Table 8. Cumulative gas and water production by field, Sand Wash Basin.[a]

Field	No. of Wells	Geologic Unit	Gas (Mcf)	Water (bbl)
Big Hole	1	Fort Union	-0-	NA
Craig Dome	16	lower Williams Fork	-0-	2,108,457
Dixon	11	lower Williams Fork	84,141	3,380,407
Lay Creek	2	lower Williams Fork Fort Union	-0-	NA
West Side Canal	6	Fort Union	-0-	>120,000[b]

[a]Cumulative to January 1, 1993, data from Petroleum Information (1993).
[b]Produced during production testing, incomplete data from operators.

well averaged 2 Mcf/d (57 m^3/d). Most of the activity was in the West Side Canal field (fig. 79), where net-coal thickness ranges from 60 to 80 ft (18 to 24 m) (Tyler and McMurry, this vol.).

Initial water production (IP) increases with permeability (Oldaker, 1991) and high water IP's (100's of bbl/d [10's of m^3/d]) are indicative of high permeability. IP's from Williams Fork coals were highest in the Yampa River valley (1,800 bbl/d [286 m^3/d]) and at the northeast margin of the basin in Dixon field, east of Baggs, Wyoming, where 1,200 bbl/d (191 m^3/d) is representative (fig. 79). The field's first well potentialed for 2,200 bbl/d (350 m^3/d). In Craig Dome field, IP's ranged from 500 to 1,000 bbl/d (80 to 159 m^3/d). IP's from Fort Union coals at West Side Canal field ranged from 100 to 3,200 bbl/d (16 to 509 m^3/d), which is a much wider range than that exhibited by Williams Fork coals at the nearby Dixon field (800 to 2,200 bbl/d [127 to 350 m^3/d]) (fig. 79). The wide Fort Union range probably reflects reservoir heterogeneity possibly due to variability in vertical flow (interconnectedness), coalbed orientation perpendicular to the lateral flow direction, and offset by faults and diagenesis along the Cherokee Arch fault system. High water potentials reflect proximity to the outcrop recharge area, basinward flow in an interconnected aquifer system, artesian conditions, and laterally extensive coal beds of high permeability. Coalbed permeability at Dixon field averages about 170 md. Because of proximity to the recharge area and high permeability, it may not be possible to economically dewater (depressure) coal beds near the basin margin. By water-well standards, coalbed methane wells are low-yield water wells; that is, they produce less than 100 gal/min (<3,430 bbl/d [<545 m^3/d]). Nevertheless, disposal costs for these volumes of water can adversely affect project economics to the extent that development may be deemed uneconomical.

Controls on Production

In the Sand Wash Basin, structural configuration, coal distribution, thermal maturity (gas content), and hydrodynamics are major controls on the occurrence and producibility of coalbed methane. Structural dip combined with topography defines the drilling-depth fairway, faults and fold axes may be sites of enhanced permeability and conventional trapping, and cleat orientation imposes permeability anisotropy. Steep structural dip and coal distribution have restricted exploration to the eastern and southeastern margins of the basin. The thickest, most laterally continuous Williams Fork coals occur in the lower part of the formation (Hamilton, this vol.). Individual coal beds are 10 to 20 ft (3 to 6 m) thick and as many as 20 beds can be present for an aggregate thickness of more than 100 ft (>30 m). The thickest, most laterally continuous Fort Union coals occur in the lower coal-bearing unit (Tyler and McMurry, this vol.). Individual coal beds are 10 to 50 ft (3 to 15 m) thick; as many as 12 beds are present for an aggregate thickness of more than 60 ft (>18 m).

Drilling-depth fairways to 6,000 and 7,500 ft (1,830 and 2,287 m) were defined by all points equal to or less than those depths derived by subtracting Williams Fork and Fort Union structural elevations from surface elevations (figs. 80 and 81). Within the fairways all Williams Fork or Fort Union coals are testable at drilling depths equal to or less than those shown. Operators and statistical analysis of depth of western coalbed completions indicate that the current economic drilling depth is approximately 6,000 ft (1,830 m). Note that the Williams Fork 6,000-ft (1,830-m) fairway hugs the eastern and southeastern margins of the basin. Rugged topography and inaccessibility will further limit development on the east. Surface elevations in the Elkhead Mountains exceed 9,000 ft (2,745 m) (Scott and Kaiser, this vol., fig. 41). The area of thickest Fort

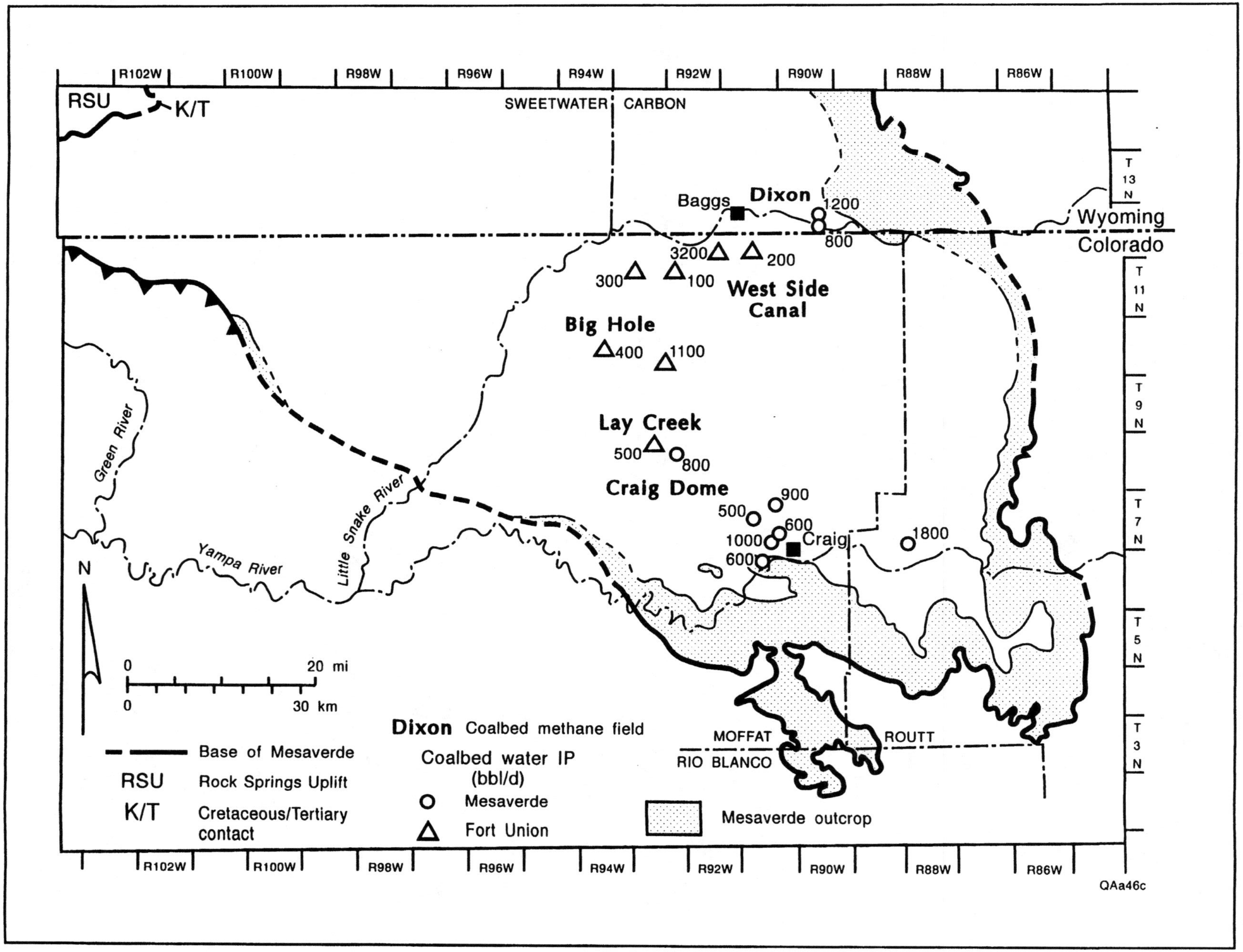

Figure 79. Initial water potentials, Williams Fork and Fort Union coals. Williams Fork IP's are high at the basin margin and reflect proximity to the recharge area and high coalbed permeability. Fort Union IP's are highest along the south-north trend of thick net coal.

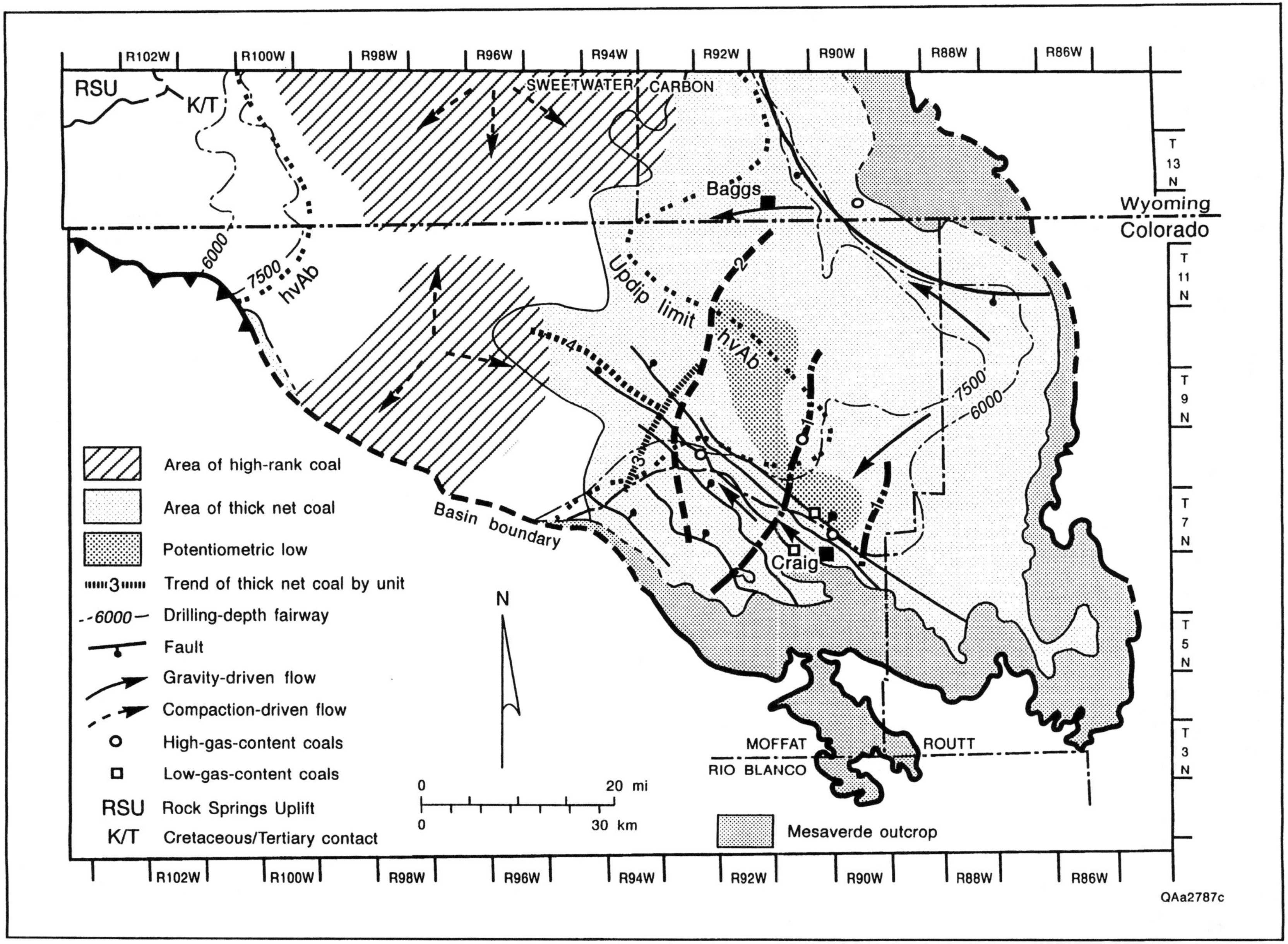

Figure 80. Geologic and hydrologic characterization of the Williams Fork Formation, Sand Wash Basin. Drilling-depth fairways hug eastern margin of the basin, reflecting steep structural dip. Ground water flows westward toward and along the Cedar Mountain fault system (CMFS) up the coal-rank gradient. Compactional flow is in the opposite direction, up and out of the basin center. Prospective coals lie basinward in association with the CMFS.

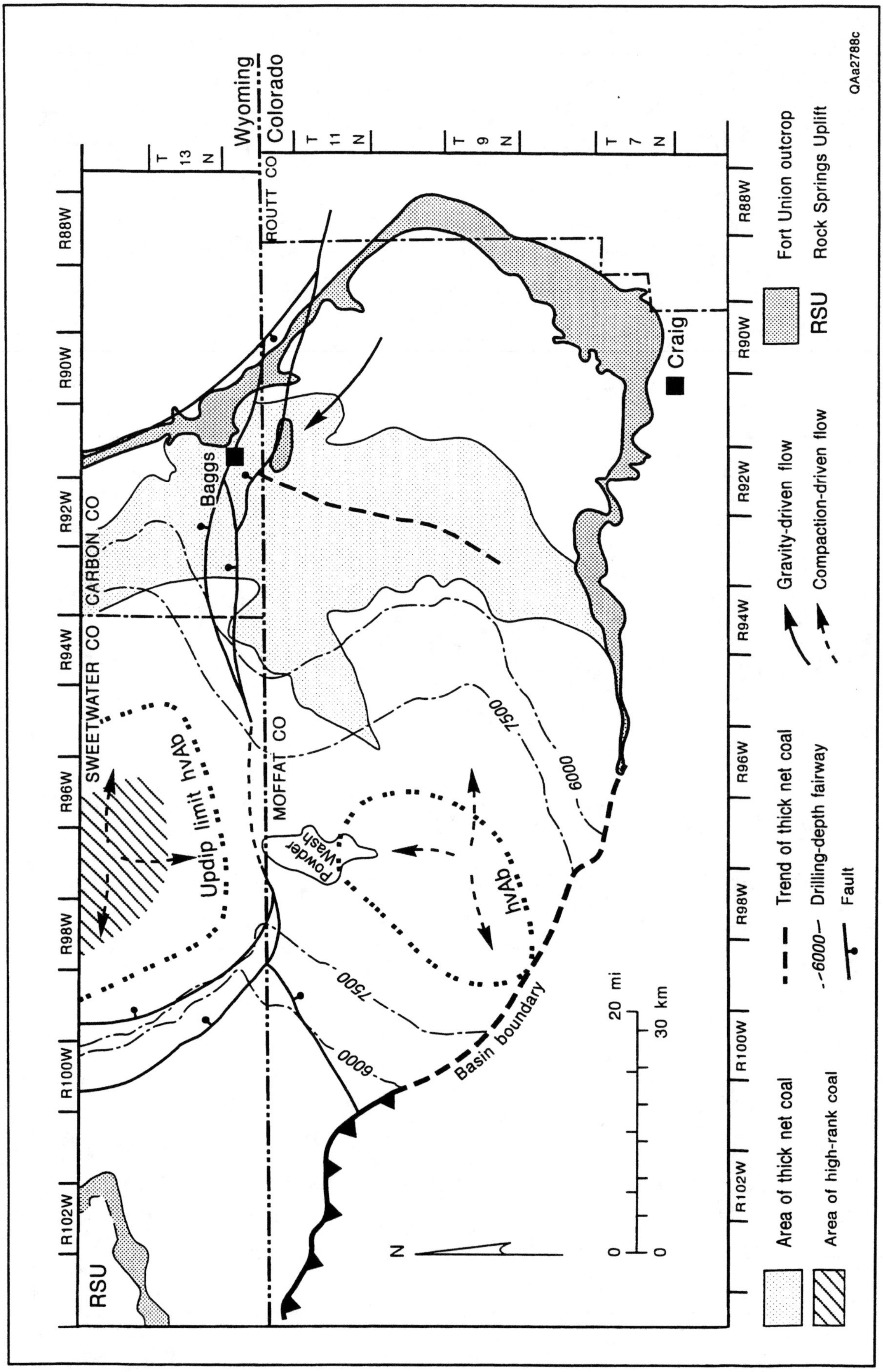

Figure 81. Geologic and hydrologic characterization of the Fort Union Formation, Sand Wash Basin. Drilling-depth fairways encompass area of thick coal. Ground water flows westward along the Cherokee Arch fault system. Compactional flow converges on the arch. Prospective coals lie in this part of the basin in the Powder Wash area.

Union coal development can be tested at depths of less than 6,000 ft (<1,830 m).

The Sand Wash Basin has no extensive area of medium-volatile bituminous and greater rank coal (Scott, this vol.), the ranks of maximum gas generation. Thus, large volumes of thermogenic gas may never have been generated from Williams Fork and Fort Union coals. Most Williams Fork coals are high-volatile C to B bituminous rank and have average gas contents of less than 200 ft^3/ton (<6.24 m^3/t) (Scott, this vol.). Most Fort Union coals are subbituminous to high-volatile C bituminous rank and have average gas contents of less than 100 ft^3/ton (<3.12 m^3/t) (Scott, this vol.). Gas contents of more than 300 ft^3/ton (>9.4 m^3/t) are generally thought necessary for commercial production (Tang and others, 1991), except where non-sorbed gas is a significant component of the production (e.g., eastern margin of Powder River Basin). Furthermore, the total production of biogenic gases in lower rank coals (less than hvAb) may not be as significant as for higher rank coals. Lower rank coals have not reached the thermal maturity level required to generate wet gases and n-alkanes, which are relatively easy for bacteria to metabolize. Although laterally extensive coals can serve as conduits for long-distance migration of gas, Williams Fork coals, except for Unit 4 coals, do not extend westward to the area of highest thermal maturity in the basin's structural center (Hamilton, this vol.) (fig. 80). Thus, they could not serve as conduits for updip, eastward, long-distance migration of gas for eventual resorption as well as possible conventional trapping.

Mesaverde ground water flows southwest and northwest, approximately parallel to net-coal trends and major fault systems, through thick coals of low thermal maturity, up the coal-rank gradient, for eventual discharge basinward (fig. 80). Consequently, only relatively small volumes of gas may be available for eventual resorption or conventional trapping downflow along potential flow barriers. Chances are best for this along the Cedar Mountain fault system, a fault zone at least 10 mi (16 km) wide and extending approximately 30 mi (~48 km) northwest and 15 mi (24 km) southeast of Craig, Colorado (Tyler and Tremain, this vol.) (fig. 80). As many as six faults are present, all downthrown to the northeast, with individual throws between 500 and 1,800 ft (152 and 549 m) for a total displacement across the system of more than 5,000 ft (>1,525 m) on top of the Mesaverde Group (Tyler and Tremain, this vol., fig. 8).

Both high- and low-gas-content Mesaverde coals occur in wells drilled in the system's fault blocks. High gas contents (>300 ft^3/ton [>9.3 m^3/t]) were reported from coals in three of five wells. Among these, one (Van Dorn 1) is near the outcrop at the south (T7N, R90W, sec. 29) and the other (Morgan 12-12) is basinward (T8N, R93W, sec. 12) (fig. 80). Gas contents were highest (>500 ft^3/ton [>15.6 m^3/t]) in coals in the Morgan well and may reflect conventional trapping (Scott, this vol., fig. 34). The third well (Blue Gravel 3-1) is located about 5 mi (8 km) northeast of the system's most northeastward fault (T8N, R91W, sec. 3). One sample from an upper Williams Fork coal bed had 432 ft^3/ton (13.48 m^3/t), but this may be an anomalous value because major coal beds below 7,000 ft (2,135 m) in the lower Williams Fork were tested but produced little gas. The two wells (Cockrell 791-4301 and Klein 23-11) with low gas contents (<100 ft^3/ton [<3.12 m^3/t]) are located near the southern outcrop (T7N, R91W, sec. 34 and T7N, R91W, sec. 11). The Cockrell 791-3401 well is close to the recharge area and is probably subject to active flow or flushing. On the other hand, the Klein 23-11 lies in a potentiometric low and is subject to regional convergent, upward flow, which should favor hydrocarbon accumulation (Tóth, 1980). However, in the absence of a seal, relatively high coalbed permeability may promote hydrocarbon flushing rather than accumulation.

There is no ready explanation for the erratic gas contents. Rank does not appear to be an important control; it is hvCb to hvAb and varies little across the eastern part of the basin. Apparently, conventional trapping and hydrodynamics are important controls. High gas content may reflect sealing along faults and less active meteoric circulation, whereas low gas content may reflect poor sealing and dynamic flow, which would promote gas loss or migration. Sealing probably depends on fortuitous juxtaposition of Williams Fork coal beds against shales of the Iles Formation. Individual faults are of insufficient throw to place Williams Fork coals against the Mancos Shale (Tyler and Tremain, this vol., fig. 8) and thus maximize the potential for conventional traps. Moreover, the Mesaverde is a regionally interconnected, leaky aquifer system with good vertical connectivity, reflecting a lack of seals and few permeability contrasts; this interconnectedness decreases the chances for conventional trapping and increases the chances for gas leakoff. Furthermore, in the absence of a regional pressure regime, gas content will be highly variable and its distribution difficult to predict. High gas contents correspond to artesian overpressure on the eastern Cherokee Arch, for example, at Dixon field (fig. 80). Finally, coal surface properties may also affect gas content. Sorption isotherms vary significantly and show at reservoir pressures low gas contents (<250 ft^3/ton [<7.8 m^3/t]) in some Williams Forks coals from widely separated wells (Scott, this vol., fig. 35).

In the Sand Wash Basin, exploration strategy is obviously to minimize water production and maximize gas content. Proximity to recharge areas should be avoided because of high possible water production. Attempted development to date has been at the basin margins, where water production is high and/or gas content low.

At Dixon field, high water production is predictable from the presence of artesian overpressure. Despite reasonable gas contents, coalbeds could not be depressurized for high productivity. At Craig Dome field, low gas contents were measured and are thought to reflect low coal rank, migration updip or along associated faults, and active recharge sweeping gas basinward entrained or in solution. Moving basinward from the recharge area should facilitate dewatering (depressuring), whereas, deeper drilling should mean higher gas content because of higher reservoir pressure and coal rank. Average Williams Fork gas content between 6,000 and 7,500 ft (1,830 and 2,287 m) is 358 ft^3/ton (11.17 m^3/t), which is more than double the value at less than 6,000 ft (<1,830 m) (table 9). Gas resources in the deeper, 1,500-ft (457-m) interval approximate those at shallow depths (<6,000 ft [<1,830 m]). Moreover, because permeability in the hydropressured section of the Sand Wash Basin is high overall, permeability adequate for commercial production may still be present at depths deeper than normally expected in western coal basins. In other words, higher overall permeability may lower the permeability floor for coalbed methane exploration in the Sand Wash Basin.

Greater emphasis should be placed on the identification of conventional traps (no-flow boundaries). Conventionally trapped gas and solution gas that can be produced with less associated water are overlooked sources of coalbed methane. Conventional traps and convergent, upward flow associated with fault zones may be favorable areas for coalbed methane exploration. The basin's gassiest coals are found in association with the Cedar Mountain fault system, a 400-mi^2 (1,036-km^2) area that has not been thoroughly tested. Most of the wells have been drilled at the fault system's south end, near the outcrop recharge area. Unit 4 Williams Fork coals at exploitable depths extend to the system's northwest end and beyond into an area of high coal rank (fig. 80). Consequently, because of migration and increased rank, gas contents may be higher, making these coals potential exploration targets. Moreover, because Unit 4 coals partially overlap underpressure, indicating limited recharge from the basin's southwest margin (Scott and Kaiser, this vol.), they may be less well connected with the regional flow system, and thus less water productive, improving chances for dewatering these coals. The Savery fault system is the western limit for Williams Fork exploration in the northeastern Sand Wash Basin. It separates hydropressure on the east from hydrocarbon overpressure to the northwest, which because of low permeability and depth restrictions is not a coalbed methane target.

Along the Cherokee Arch, considerable Fort Union conventional gas and oil production has been established (Barlow and others, 1993; Mullen and Tremain, 1993) that probably reflects convergent, upward flow as well as structural and stratigraphic trapping. Tóth (1980) argues that areas of converging and ascending flow favor the accumulation of hydrocarbons. Upward leakage and/or migration updip of water and hydrocarbons from the deeper parts of the basin along the Cherokee Arch fault system and flanks of the arch is inferred from a flattened Fort Union potentiometric surface (Scott and Kaiser, this vol.). Although Fort Union coals are thinner and less numerous and still of low rank westward along the Cherokee Arch and into the Powder Wash field area (Tyler and McMurry, this vol.; Scott, this vol.), they may be highly charged with gas and thus good candidates for completion in the course of conventional gas development. In fact, BHP Petroleum has recently proposed to workover a Powder Wash field well (T11N, R97W, sec. 17) by perforating three Fort Union coals (Petroleum Information, 1993) (fig. 81).

Producibility

On the basis of studies in the San Juan Basin (Kaiser and others, 1991a) and herein in the Sand Wash Basin, geologic and hydrologic controls critical to the producibility of coalbed methane have been identified. However, simply knowing what those controls are will not lead to a conclusion about producibility. It is the interplay of several geologic and hydrologic controls and their spatial relation that governs producibility (fig. 82). For example, permeability that is too high results in high water production and is as detrimental to the production of coalbed methane as very low permeability. Depositional and structural setting control the distribution of the coal reservoirs and determine their location with respect to the thermally mature parts of the basin and their orientation relative to the direction of ground-water flow. Flow may be perpendicular or parallel to the structural grain or up or down the coal-rank gradient. Gas content reflects not only coal rank but also permeability contrasts and ground-water flow, as permeability and flow influence conventional trapping of gas, reservoir pressure (which is commonly hydrodynamically controlled), and migration of gas. High productivity requires that geologic and hydrologic controls be synergistically combined. This synergism is evident in a comparison of the San Juan and Sand Wash Basins, which are thought to represent end members of a producibility continuum. Although the basins are end members with respect to production, they do not lie at opposite ends of the continuum for key controls (fig. 83). In fact, they share controls that overlap, yet they have widely disparate productivities.

Geologic and hydrologic comparison of the San Juan and Sand Wash Basins shows fundamental differences

Table 9. Coal and gas resources in the Sand Wash Basin based on depth to the base of the coal-bearing unit.

	Resources		Percent of total resource		Average gas content scf/ton
	Coal (billion tons)	Gas (Tcf)	Coal	Gas	
Sand Wash Basin					
< 6,000 ft	180	24	48	23	133
6,000 to 7,500 ft	67	20	18	20	299
> 7,500 ft	130	58	34	57	446
Williams Fork Formation					
< 6,000 ft	122	21	44	21	172
6,000 to 7,500 ft	53	19	19	21	358
> 7,500 ft	105	54	37	58	524
Fort Union Formation					
< 6,000 ft	58	3	60	43	52
6,000 to 7,500 ft	14	1	14	14	71
> 7,500 ft	25	3	26	43	120

between them that explain prolific and marginal production, respectively, in the two basins (fig. 84). In the San Juan Basin, ground water flows down the coal-rank gradient from the northern basin margin through thick coals of high rank orthogonally toward lower rank coals at a structural hingeline (no-flow boundary) along which high coalbed methane production occurs. High coal rank suggests that relatively large volumes of gas are potentially available to be entrained or dissolved and swept basinward ahead of an advancing flux of meteoric water for eventual resorption and conventional trapping along the hingeline. Flow turns steeply upward at this point upon pinch-out of thick aquifer coal beds and/or their offset by faults along the hingeline (fig. 85). Northeast of this no-flow boundary (permeability barrier), appreciable conventional free gas and solution gas, in addition to that sorbed on the coal surface, are thought to be present (Kaiser and Ayers, 1991; Kaiser and others, 1991a). Contribution from nonsorbed gas conventionally trapped and concentrated at the hingeline, and high coalbed permeability, explain exceptionally high production at this point in the San Juan Basin. In the Sand Wash Basin, ground water flows westward from an eastern recharge area through low-rank, low-gas-content coals up the coal-rank gradient toward areas of higher thermal maturity. Consequently, only relatively small volumes of gas may be available for eventual resorption and conventional trapping basinward along potential flow barriers such as the leaky Cedar Mountain fault system.

In summary, the essential elements of producibility are (1) thick, laterally continuous coals of high thermal maturity, (2) basinward flow of ground water through high-rank, high-gas-content coals down the coal-rank gradient toward no-flow boundaries (structural hingelines, faults, facies changes, discharge areas) oriented perpendicular to the regional flow direction, and (3) conventional trapping of gas along those boundaries. When flow boundaries and flow direction are perpendicular, the largest possible area of flow is intercepted, maximizing gas accumulation. Conventional trapping provides additional sources of gas beyond that sorbed on the coal surface.

Conclusions

1. Total gas resources in the Sand Wash Basin are large (101 Tcf [2.86 Tm^3]) because total coal resources are large (377 billion tons [342 billion t]). Gas and coal resources at shallow drilling depths of less than 6,000 ft (<1,830 m) are 24 Tcf (680 Bm^3) and 180 billion tons (163 billion t), respectively. More than 90 percent of the total gas resources and 74 percent of the total coal resources are in the Williams Fork Formation.

2. Coalbed wells to date have yielded little gas and large volumes of water. The basin's cumulative gas/water ratio is approximately 15 ft^3/bbl (~2.7 m^3/m^3).

3. To date, low gas content (<100 to 200 ft^3/ton [<3.12 to 6.24 m^3/t]) and high water production (100's of bbl/d [10's of m^3/d]) have limited coalbed methane activity in the basin. Low gas contents in areas

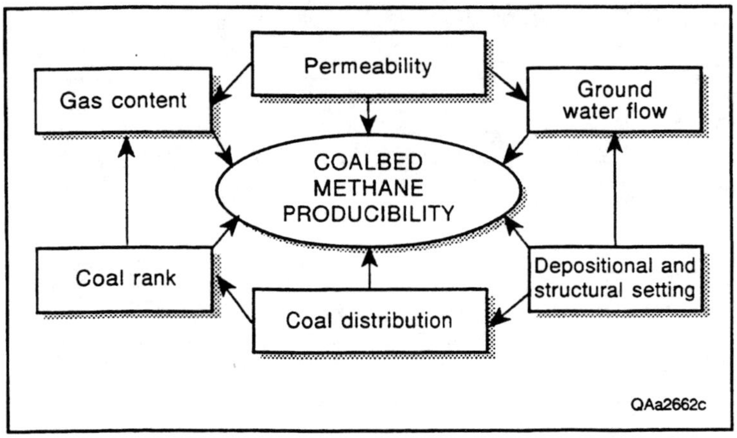

Figure 82. Interplay of geologic and hydrologic controls governs the producibility of coalbed methane.

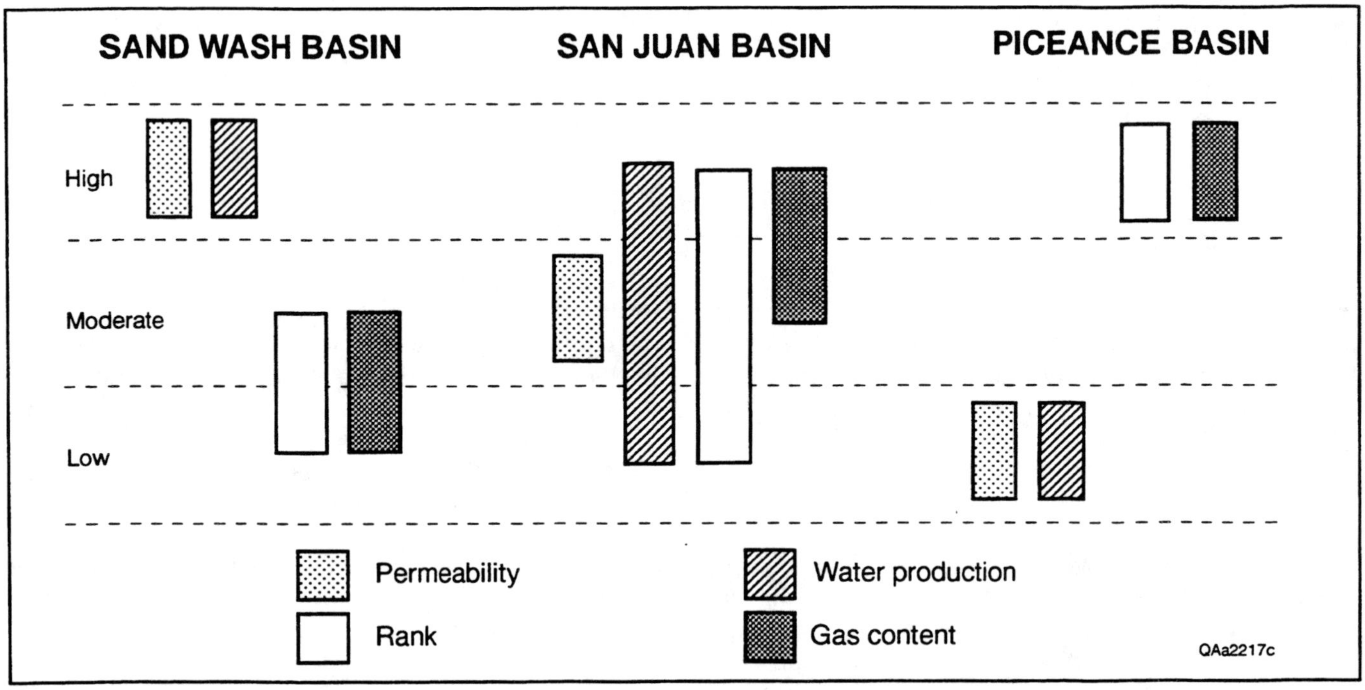

Figure 83. Characteristics of key geologic and hydrologic controls overlap in western coal basins.

drilled reflect (a) nondeposition of coals in the basin's most thermally mature area, which reduces the potential for long-distance migration of gas, (b) low coal rank (below that for maximum gas generation) at drilled total depths, (c) hydrodynamics such that flow up the coal-rank gradient and good aquifer interconnectedness reduce the chances for conventional trapping of gas, and (d) coal surface properties. High water production reflects regional aquifer systems of high transmissivity.

4. Steep structural dip and coal distribution have restricted exploration to the eastern and southeastern margins of the basin, where Williams Fork coals were the prime coalbed methane targets. The most prospective coals lie basinward. Gas contents in some Williams Fork coals associated with the Cedar Mountain fault system northwest of Craig, Colorado, exceed 400 ft^3/ton (12.5 m^3/t). Fort Union coals, though low rank, may be highly charged with gas westward along the Cherokee Arch and into the Powder Wash field area, due to migration of gas up and out of the basin toward the arch.

5. High productivity requires that geologic and hydrologic controls on production (permeability, ground-water flow direction, coal distribution and rank, gas content, and depositional and structural setting) be synergistically combined. That synergism is evident in a comparison of the San Juan and Sand Wash Basins. In

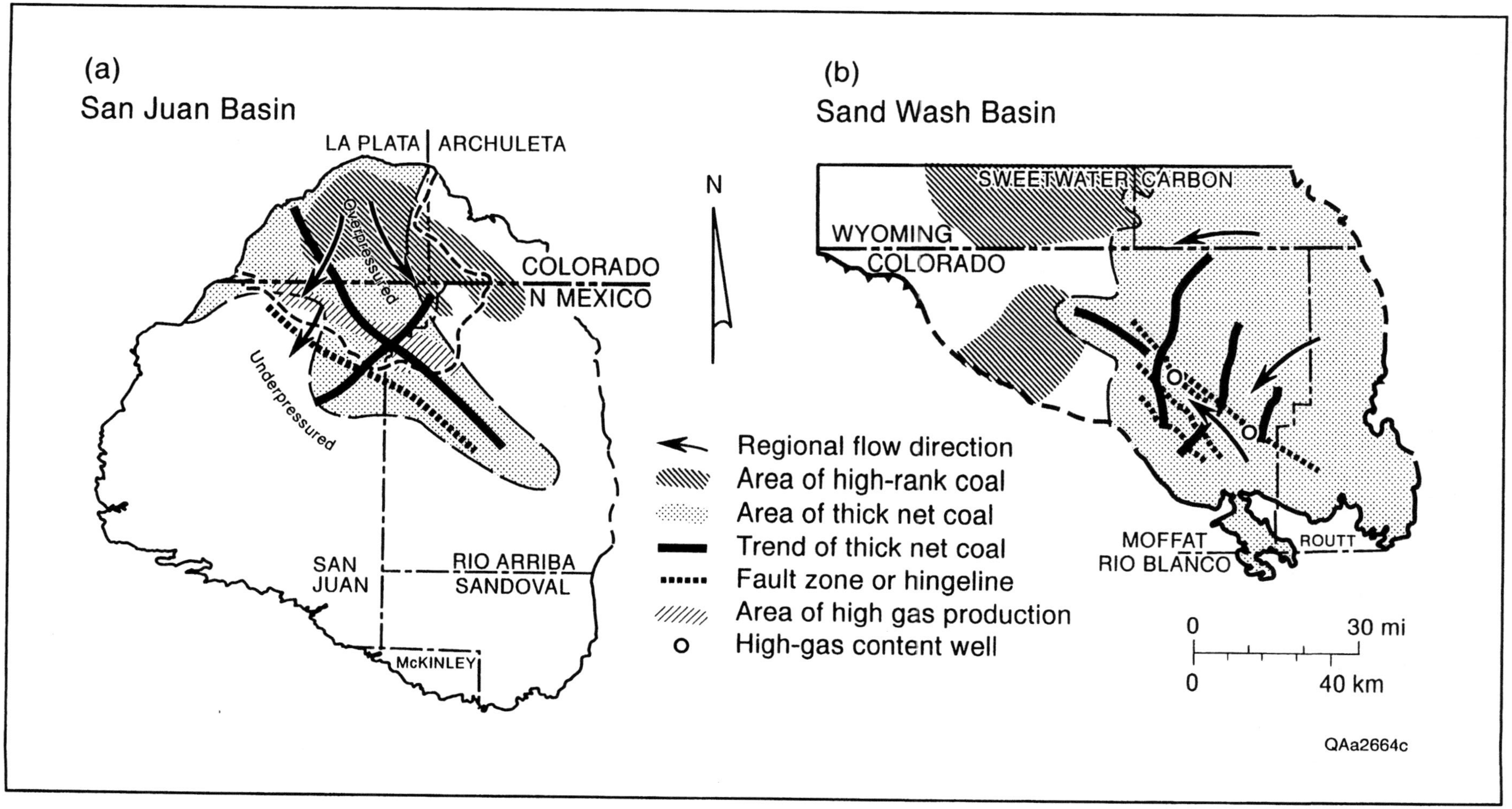

Figure 84. Geologic and hydrologic comparison of the San Juan and Sand Wash Basins. In the San Juan Basin, ground water flows through thick coals of high rank toward a structural hingeline (no flow boundary) along which prolific production occurs. In the Sand Wash Basin, flow is through coals of low thermal maturity toward a leaky, regional fault system along which high- and low-gas-content coals occur.

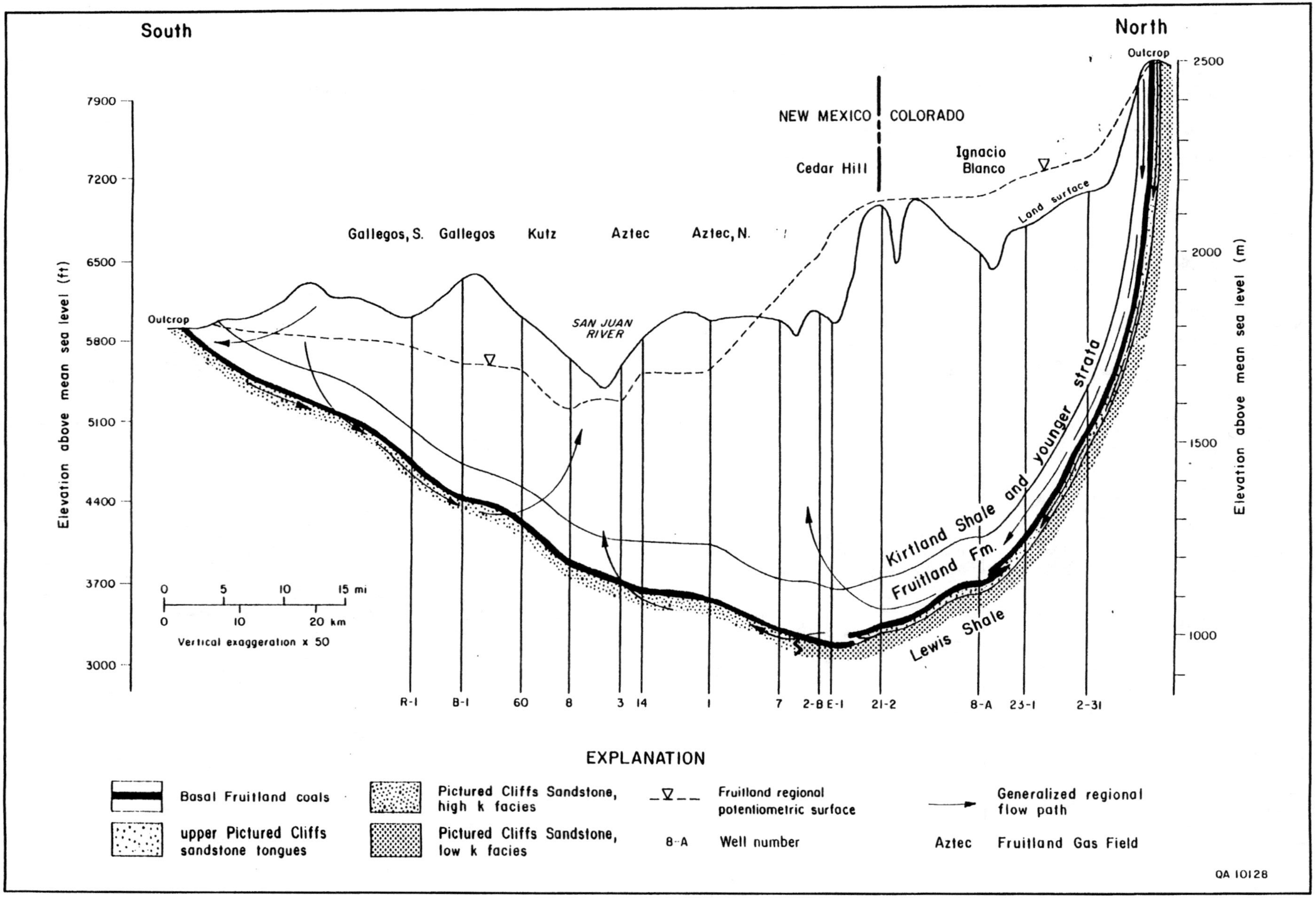

Figure 85. Schematic cross-sectional ground-water flow, San Juan Basin (modified from Kaiser and others, 1991b). Flow is down the steep northern limb from high to lower rank coal. Flow turns steeply upward in the vicinity of Cedar Hill field, upon pinch-out of thick coal beds and/or their offset by faults, along a structural hingeline. Free gas and solution gas, conventionally trapped and concentrated at a no-flow boundary, and high permeability combine for prolific production at this point in the basin.

the San Juan Basin, ground water flows basinward through high-rank, high-gas-content coals, down the coal-rank gradient orthogonally toward a structural hingeline (no-flow boundary) along which exceptionally high production occurs. A giant conventional-hydrodynamic trap is postulated along the hingeline and implies that non-sorbed, free, conventionally trapped gas and solution gas are overlooked sources of coalbed methane. In the Sand Wash Basin, flow is basinward through low-rank, low-gas-content coals, up the coal-rank gradient toward a major fault system along which dynamic flow occurs. Consequently, only small volumes of gas are available for eventual resorption and possible conventional trapping downflow along a leaky fault system.

6. A basin-scale coalbed methane producibility model is evolving out of the comparison between the San Juan and Sand Wash Basins; its essential elements are ground-water flow through thick coals of high rank and high gas content orthogonally toward no-flow boundaries and conventional trapping of gas along them. The model remains to be tested and refined in other coal basins.

Acknowledgments

This research was funded by the Gas Research Institute under contract no. 5091-214-2261, Richard A. McBane and Dave Hill, project managers. Industry data greatly strengthened the study and in particular facilitated analysis of thermal history and production. We appreciate the time and effort required to assemble and provide these data, and we thank all those involved. Meridian Oil, Inc., Fuel Resources Development Co. (Fuelco), Quintana Petroleum Corporation, Chevron U.S.A., Inc., and others provided gas-content and vitrinite-reflectance data. Meridian, Fuelco, Cockrell Oil Corporation, and Quintana provided production-test data and patiently answered questions about their coalbed wells' production histories. Operators also provided extensive suites of geophysical logs for many of their coalbed wells. Without operator cooperation, our understanding of the interplay among gas content, coal rank, production, and hydrogeology would be sketchy. Discussions with Curtis Johnson, John Wessels, Tom Bergstresser, Tom Daneker, and Ben Weichman were particularly instructive. Dwight's Energydata, Inc., supplied initial potentials and chemical analyses of produced formation waters. Union Pacific Resources contributed seismic data used to better define the Cedar Mountain fault system.

Joseph Yeh's assistance in the implementation of computer programs to sort DST data and to calculate resources was invaluable. S. E. Laubach was a ready source of structural expertise and a reviewer of the study's structural geology. R. G. McMurry, L. T. Ortiz, L. M. Scott, and Naijiang Zhou compiled structural, cleat, fault, and fold data; Naijiang Zhou also compiled DST, gas-content, and stratigraphic data. Drafting was by the Bureau's cartographic staff under the direction of R. L. Dillon, chief cartographer. Word processing was by Susan Lloyd, editing was by Amanda R. Masterson, and design by Diane Ruetz.

REFERENCES

American Society for Testing and Materials (ASTM), 1983, Standard classification of coals by rank; ASTM designation D388-82 in gaseous fuels, coal and coke: Philadelphia, ASTM, 1983 Book of Standards, v. 0505, 5 p.

Ammosov, I. I., and Eremin, I. V., 1960, Fracturing in coal: Moscow, IZDAT Publishers (translated from Russian), 109 p. (Available from the Office of Technical Services, Washington, D.C.)

Asquith, D. O., 1970, Depositional topography and major marine environments, Late Cretaceous, Wyoming: American Association of Petroleum Geologists Bulletin, v. 54, no. 7, p. 1184–1224.

Barlow, J. A., Mullen, D. M., and Tremain, C. M., 1993, Paleocene Fort Union Formation, *in* Atlas of major Rocky Mountain gas reservoirs: New Mexico Bureau of Mines and Mineral Resources, p. 36–37.

Barrs, D. L., Bartleson, B. L., Chapin, C. E., Curtis, B. F., De Voto, R. H., Everett, J. R., Johnson, R. C., Molenaar, C. M., Peterson, F., Schenk, C. J., Love, J. D., Merin, I. S., Rose, P. R., Ryder, R. T., Waechter, N. B., and Woodward, L. A., 1988, Basins of the Rocky Mountain Region, *in* Sloss, L. L., ed., Sedimentary cover—North American craton, U.S.: Geological Society of America, Decade of North American Geology, v. D-2, p. 109–220.

Bass, N. W., Eby, J. B., and Campbell, M. R., 1955, Geology and mineral fuels of parts of Routt and Moffat Counties, Colorado: U.S. Geological Survey Bulletin 1027-D, 250 p.

Beaumont, E. A., 1979, Depositional environments of Fort Union sediments (Tertiary, northwest Colorado) and their relation to coal: American Association of Petroleum Geologists Bulletin, v. 63, no. 2, p. 194–217.

Beck, R. A., Vondra, C. F., Filkins, J. E., and Olander, J. D., 1988, Syntectonic sedimentation and Laramide basement thrusting, Cordilleran foreland; timing of deformation, *in* Schmidt, C. J., ed., Interaction of the Rocky Mountain Foreland and the Cordilleran Thrust Belt: Geological Society of America, Memoir 171, p. 465–487.

Boreck, D. L., Tremain, C. M., Sitowitz, Linda, and Lorenson, T. D., 1981, The coalbed methane potential of the Sand Wash Basin, Green River coal region, Colorado: Colorado Geological Survey Open-File Report 81-6, 25 p.

Boreck, D. L., Jones, D. C., Murray, D. K., Schultz, J. E., and Suek, D. C., 1977, Colorado coal analyses, 1975 (analyses of 64 samples collected in 1975): Colorado Geological Survey Information Series 7, 112 p.

Boyles, J. M., and Scott, A. J., 1981, Depositional systems Upper Cretaceous Mancos Shale and Mesaverde Group, northwestern Colorado, *in* Boyles, J. M., Kauffman, E. G., Kiteley, L. W., and Scott, A. J., Depositional systems Upper Cretaceous Mancos Shale and Mesaverde Group, northwestern Colorado: Rocky Mountain Section, Society of Economic Paleontologists and Mineralogists, field trip guidebook, part 1, 82 p.

Collentine, M., Libra, R., Feathers, K. R., and Hamden, L., 1981, Occurrence and characteristics of ground water in the Great Divide and Washakie Basins, Wyoming: Laramie, University of Wyoming Water Resources Research Institute, report prepared for U.S. Environmental Protection Agency under contract no. G-008269-79, v. VI-A, 112 p., 5 appendices.

Colorado Climate Center, 1984, Colorado average annual precipitation 1951–1980: Fort Collins, Colorado State University, Department of Atmospheric Science, Colorado Climate Center, scale 1:500,000.

Colson, C. T., 1969, Stratigraphy and production of the Tertiary formations in the Sand Wash and Washakie Basins, *in* Barlow, J. A., Jr., ed., Tertiary rocks of Wyoming: Wyoming Geological Association Guidebook, 21st field conference, p. 121–128.

Creedy, D. P., 1988, Geologic controls on the formation and distribution of gas in British coal measurement strata: International Journal of Coal Geology, v. 10, no. 1, p. 1–31.

Cronoble, J. M., 1969, South Baggs–West Side Canal gas field, Carbon County, Wyoming and Moffat County, Colorado, *in* Barlow, J. A., Jr., ed., Tertiary rocks of Wyoming: Wyoming Geological Association Guidebook, 21st field conference, p. 129–137.

Diamond, W. P., and Levine, J. R., 1981, Direct method determination of the gas content of coal—procedures and results: U.S. Bureau of Mines Report of Investigations 8515, 36 p.

Edmunds, W. M., 1981, Hydrogeochemical investigations, *in* Lloyd, J. W., ed., Case studies in ground water resources evaluation: New York, Oxford University Press, p. 87–112.

Ettinger, I., Eremin, I., Zimakov, B., and Yanovskaya, M., 1966, Natural factors influencing coal sorption properties I—petrography and the sorption properties of coals: Fuel, v. 45, p. 267–282.

Frazier, D. E., 1974, Depositional episodes: their relationship to the Quaternary stratigraphic framework in the northwestern portion of the Gulf Basin: The University of Texas at Austin, Bureau of Economic Geology Geological Circular 74-1, 28 p.

Galloway, W. E., 1989, Genetic stratigraphic sequences in basin analysis 1: architecture and genesis of flooding surface bounded depositional units: American Association of Petroleum Geologists Bulletin, v. 73, p. 125–142.

Galloway, W. E., and Hobday, D. K., 1983, Terrigenous clastic depositional systems: applications to petroleum, coal, and uranium exploration: New York, Springer-Verlag, 423 p.

Gill, J. R., Merewether, E. A., and Cobban, W. A., 1970, Stratigraphy and nomenclature of some Upper Cretaceous and lower Tertiary rocks in south-central Wyoming: U.S. Geological Survey Professional Paper 667, 53 p.

Glass, G. B., 1981, Coal deposits of Wyoming, *in* Epis, R. C., and Callender, J. F., eds., Western Slope Colorado: New Mexico Geological Society Guidebook, 32nd field conference, p. 181–236.

Gries, Robbie, 1983, Oil and gas prospecting beneath Precambrian of foreland thrust plates in Rocky Mountains: American Association of Petroleum Geologists Bulletin, v. 67, p. 1–28.

Grout, M. A., and Verbeek, E. R., 1992a, Fracture history of the Divide Creek and Wolf Creek anticlines and its relation to Laramide Basin—margin tectonism, southern Piceance Basin, northwestern Colorado: U.S. Geological Survey Bulletin 1787-Z, 32 p.

_________ 1992b, Joint-history summary and orientation data for Upper Cretaceous sandstones, Rock Springs and Rawlins uplifts, Washakie Basin, southern Wyoming: U.S. Geological Survey Open-File Report 92-338, 40 p.

Hancock, E. T., 1925, Geology and coal resources of the Axial and Monument Butte quadrangles, Moffat County, Colorado: U.S. Geological Survey Bulletin 757, 134 p.

Hansen, D. E., 1986, History of faulting in the eastern Uinta Mountains, Colorado and Utah, *in* Stone, D. S., ed., New interpretations of northwest Colorado geology: Denver, Rocky Mountain Association of Geologists, p. 5–17.

Hansen, W. R., 1965, Geology of the Flaming Gorge area, Utah–Colorado–Wyoming: U.S. Geological Survey Professional Paper 490, 196 p.

Haun, J. D., 1961, Stratigraphy of post-Mesaverde Cretaceous rocks, Sand Wash Basin and vicinity, Colorado and Wyoming, *in* Wiloth, G. J., and others, eds., Late Cretaceous rocks—Green River, Washakie, Wind River and Powder River Basins: Wyoming Geological Association Guidebook, 16th field conference, p. 116–124.

Haun, J. D., and Weimer, R. J., 1960, Cretaceous stratigraphy of Colorado, *in* Weimer R. J., and Haun, J. D., eds., Guide to the geology of Colorado: Geological Society of America, Rocky Mountain Association of Geologists and Colorado Scientific Society Guidebook, p. 58–65.

Heller, P. L., Bowdler, S. S., Chambers, H. P., Coogan, J. C., Hagen, E. S., Shuster, M. W., Winslow, N. S., and Lawton, T. F., 1986, Time of initial thrusting in the Sevier orogenic belt, Idaho, Wyoming, and Utah: Geology, v. 14, no. 5, p. 388–391.

Hettinger, R. D., Honey, J. G., and Nichols, D. J., 1991, Chart showing correlations of Upper Cretaceous Fox Hills sandstone and Lance Formation, and Lower Tertiary Fort Union, Wasatch, and Green River Formations, from the eastern flank of the Washakie Basin to the southeastern part of the Great Divide Basin, Wyoming: U.S. Geological Survey Miscellaneous Investigations Series, Map I-2151.

Hettinger, R. D., and Kirschbaum, M. A., 1991, Chart showing correlations of some Upper Cretaceous and Lower Tertiary rocks, from the east flank of the Washakie Basin to the east flank of the Rock Springs Uplift, Wyoming: U.S. Geological Survey Miscellaneous Investigations Series, Map I-2152.

Honey, J. G., and Hettinger, R. D., 1989, Stratigraphic sections showing coal correlations within the lower part of the Fort Union Formation, Fillmore Ranch and Seaverson Reservoir Quadrangles, Carbon County, Wyoming: U.S. Geological Survey Coal Investigations Map C-0127.

Honey, J. G., and Roberts, L. N. R., 1989, Stratigraphic sections showing coal correlations within the lower part of the Fort Union Formation in the Baggs area, Carbon County, Wyoming: U.S. Geological Survey Coal Investigations Map, Map C-135, scale 1:100,000.

Hunt, J. M., 1979, Petroleum geochemistry and geology: San Francisco, W. H. Freeman and Company, 617 p.

Hutton, A. C., and Cook, A. C., 1980, Influence of alginite on the reflectance of vitrinite from Joadja, N. S. W., and some other coals and oil shales containing alginite: Fuel, v. 59, p. 711–714.

Irving, E., 1979, Paleopoles and paleolatitudes of North America and speculations about displacement terrains: Canadian Journal of Earth Sciences, v. 16, p. 669–694.

Irwin, C. D., 1986, Upper Cretaceous and Tertiary cross sections, Moffat County, Colorado, *in* Stone, D. S., ed., New interpretations of northwest Colorado geology: Denver, Rocky Mountain Association of Geologists, p. 151–156.

James, T. A., and Burns, B. J., 1984, Microbial alteration of subsurface natural gas accumulations: American Association of Petroleum Geologists Bulletin, v. 68, p. 957–960.

Jenden, P. D., 1985, Analysis of gases in the Earth's crust: Gas Research Institute, final report, GRI-85/0106 (January 1982–February 1985), 191 p.

Johnson, R. C., and Nuccio, V. F., 1986, Structural and thermal history of the Piceançe Creek Basin, western Colorado, in relation to hydrocarbon occurrence in the Mesaverde Group, *in* Spencer, C. W., and Mast, R. F., eds., Geology of tight gas reservoirs: American Association of Petroleum Geologists Studies in Geology 24, p. 165–205.

Joubert, J. I., Grein, C. T., and Bienstock, Daniel, 1973, Sorption of methane in moist coal: Fuel, v. 52, p. 181–185.

_________ 1974, Effect of moisture on the methane capacity of American coals coal: Fuel, v. 53, p. 186–191.

Juntgen, H., and Karweil, J., 1966, Gasbildung und gasspeicherung in steinkohlenflozen, parts I and II: Erdol and Kohle, Erdgas, Petrochemie, v. 19, p. 251–258 and 339–334.

Kaiser, W. R., 1993, Abnormal pressure in coal basins of the western United States, *in* Proceedings of the 1993 International Coalbed Methane Symposium, The University of Alabama, School of Mines and Energy Development, paper 9333, v. 1, p. 174–186.

Kaiser, W. R., and Ayers, W. B., Jr., 1991, Geologic and hydrologic characterization of coalbed methane reservoirs, Fruitland Formation, San Juan Basin, Colorado and New Mexico: Richardson, Texas, Society of Petroleum Engineers, SPE paper 23458, 14 p.

Kaiser, W. R., Ayers, W. B., Jr., Ambrose, W. A., Laubach, S. E., Scott, A. R., and Tremain, C. M., 1991a, Geologic and hydrologic characterization of coalbed methane production, Fruitland Formation, San Juan Basin, *in* Ayers, W. B., Jr., and others, Geologic and hydrologic controls on the occurrence and producibility of coalbed methane, Fruitland Formation, San Juan Basin: Chicago, Gas Research Institute topical report GRI-91/0072, p. 273–301.

Kaiser, W. R., Swartz, T. E., and Hawkins, G. J., 1991b, Hydrology of the Fruitland Formation, San Juan Basin, *in* Ayers, W. B., Jr., and others, Geologic and hydrologic controls on the occurrence and producibility of coalbed methane, Fruitland Formation, San Juan Basin: Chicago, Gas Research Institute topical report GRI-91/0072, p. 195–241.

Kalkreuth, W. D., 1982, Rank and petrographic composition of selected Jurassic–Lower Cretaceous coals of British Columbia, Canada: Bulletin of Canadian Petroleum Geology, v. 30, no. 2, p. 112–139.

Kauffman, E. G., 1977, Geological and biological overview—Western Interior Cretaceous basin, *in* Kauffman, E. G., ed., Cretaceous facies, faunas, and paleoenvironments across the Western Interior basin: The Mountain Geologist, v. 6, p. 227–245.

Khalsa, N. S., and Ladwig, L. R., 1981, Colorado coal analyses 1976–1979: Colorado Geological Survey Information Series 10, 364 p.

Laubach, S. E., Tyler, Roger, Ambrose, W. A., and Tremain, C. M., 1992a, Preliminary map of fracture patterns in coal in the western United States, *in* Fractured and jointed rock masses: International Society for Rock Mechanics: prepared for the U.S. Department of Energy under contract no. DE-ACOC-765F00098, v. 1, p. 183–190.

_________ 1992b, Preliminary map of fracture patterns in

coal in the Western United States, *in* Fractured and jointed rock masses: Lake Tahoe, California, Granlibakken Conference Center, preprints, v. 1, p. 183–190.

Laubach, S. E., Tyler, Roger, Ambrose, W. A., Tremain, C. M., and Grout, M. A., 1992c, Preliminary map of fracture patterns in coal in the western United States, *in* Mullen, C. E., ed., Rediscover the Rockies: Wyoming Geological Association Guidebook, 45th field conference, p. 253–267.

Law, B. E., 1984, Relationships of source-rock, thermal maturity, and overpressuring to gas generation and occurrence in low-permeability Upper Cretaceous and low Tertiary rocks, Greater Green River Basin, Wyoming, Colorado, and Utah, *in* Woodward, Jane, Meissner, F. F., and Clayton, J. L., eds., Hydrocarbon source rocks of the greater Rocky Mountain region: Denver, Rocky Mountain Association of Geologists, p. 469–490.

_________ 1992, Thermal maturity patterns of Cretaceous and Tertiary rocks, San Juan Basin, Colorado and New Mexico: Geological Society of America Bulletin, v. 103, p. 192–207.

Law, B. E., and Dickinson, W. W., 1985, Conceptual model for origin of abnormally pressured gas accumulations in low-permeability reservoirs: American Association of Petroleum Geologists Bulletin, v. 69, no. 8, p. 1295–1304.

Law, B. E., Pollastro, R. M., and Keighin, C. W., 1986, Geologic characterization of low-permeability gas reservoirs in selected wells, Greater Green River Basin, Wyoming, Colorado, and Utah, *in* Spencer, C. W., and Mast, R. F., eds., Geology of tight gas reservoirs: American Association of Petroleum Geologists Studies in Geology 24, p. 253–269.

Law, B. E., Rice, D. D., and Flores, R. M., 1991, Coalbed gas accumulations in the Paleocene Fort Union Formation, Powder River Basin, Wyoming, *in* Schwochow, S. D., Murray, D. K., and Fahy, M. F., eds., Coalbed methane of western North America: Denver, Rocky Mountain Association of Geologists, p. 179–190.

Law, B. E., Spencer, C. W., Charpentier, R. R., Crovelli, R. A., Mast, R. F., Dolton, G. L., and Wandrey, C. J., 1989, Estimates of gas resources in overpressured low-permeability Cretaceous and Tertiary sandstone reservoirs, Greater Green River Basin, Wyoming, Colorado, and Utah, *in* Eisert, J. L., ed., Gas resources of Wyoming: Wyoming Geological Association Guidebook, 40th field conference, p. 39–61.

Levine, J. R., 1987, Influence of coal composition on the generation and retention of coalbed natural gas: Proceedings of the 1987 coalbed methane symposium, University of Alabama, School of Mines and Energy Development, p. 15–18.

_________ 1992, An overview of coal composition, *in* Ayers, W. B., Jr., Kaiser, W. R., and Levine, J. R., Coalbed methane: depositional, hydrologic, and petrologic controls on reservoir characteristics of coal beds: American Association of Petroleum Geologists/Energy Minerals Division, Short Course No. 13, presented at the 1992 Annual Convention, Calgary, Alberta, Canada, p. 58–160.

_________ in press, Coalification: the evolution of coal as a source rock and reservoir rock for oil and gas, *in* Law, B. E., ed., Hydrocarbons from coal: American Association of Petroleum Geologists Special Publication.

Lickus, M. R., and Law, B. E., 1988, Structure contour map of the Greater Green River Basin, Wyoming, Colorado, and Utah: U.S. Geological Survey Miscellaneous Field Studies Map MF-2031.

Livesey, G. B., 1985, Laramide structures of the southeastern Sand Wash Basin, *in* Gries, R. R., and Dyer, R. C., eds., Seismic exploration of the Rocky Mountain region: Rocky Mountain Association of Geologists and Denver Geophysical Society, p. 87–94.

MacGowan, D. B., and Britton, Douglas, 1992, Organic geochemistry and maturation trends of shales and coals of the Almond Formation, *in* Surdam, R. C., and others, Natural gas resource characterization study of the Mesaverde Group in the Greater Green River Basin, Wyoming: a strategic plan for the exploration of tight gas sands: Chicago, Gas Research Institute Annual Report, GRI-93/0112, p. 53–60.

Mahajan, O. P., 1989, Adsorption and pore structure in coal-water interactions, *in* Klein, R., and Wellek, R., eds., Sample selection, aging, and reactivity of coal: New York, John Wiley Interscience, p. 157–214.

Masters, C. D., 1961, Fort Union Formation—eastern Sand Wash Basin, Colorado, *in* Wiloth, G. J., and others, eds., Late Cretaceous rocks—Green River, Washakie, Wind River, and Powder River Basins: Wyoming Geological Association Guidebook, 16th field conference, p. 125–128.

Mavor, M. J., Dhir, Rahul, McLennan, J. D., and Close, J. C., 1991, Evaluation of the hydraulic fracture stimulation of the Colorado 32-7 No. 9 well, San Juan Basin, *in* Schwochow, S. D., Murray, D. K., and Fahy, M. F., eds., Coalbed methane of western North America: Rocky Mountain Association of Geologists, fall conference and field trip guidebook, p. 241–248.

McDonald, R. E., 1972, Paleocene and Eocene rocks of the central and southern Rocky Mountain basins, *in* Mallory, W. W., ed., Geologic atlas of the Rocky Mountain region: Denver, Rocky Mountain Association of Geologists, p. 243–256.

_________ 1975, Structure, correlation and depositional environments of the Tertiary, Sand Wash and Washakie Basins, Colorado and Wyoming, *in* Bolyard, D. W., ed., Deep drilling frontiers of the central Rocky Mountains: Denver, Rocky Mountain Association of Geologists, p. 175–184.

McPeek, L. A., 1981, Eastern Green River Basin: a developing giant gas supply from deep, overpressured Upper Cretaceous sandstones: American Association of Petroleum Geologists Bulletin, v. 65, no. 6, p. 1078–1098.

Moffat, D. H., and Weale, K. E., 1955, Sorption by coal of methane at high pressures: Fuel, v. 34, p. 449–462.

Mountain Fuel Supply Company, 1961, Sugar Loaf field, Almond Pool, *in* Parker, J. M., ed., Colorado–Nebraska oil and gas fields: Denver, Rocky Mountain Association of Geologists, p. 238.

Mullen, D. M., and Tremain, C. M., 1993, Eocene Wind River and Wasatch Formations, *in* Atlas of major Rocky Mountain gas reservoirs: New Mexico Bureau of Mines and Mineral Resources, p. 34–35.

Murray, D. K., Fender, H. B., and Jones, D. C., 1977, Coal and methane gas in the southeastern Piceance Creek Basin, Colorado, *in* Veal, H. K., ed., Exploration frontier of the central and southern Rockies, Rocky Mountain Association Geological Field Conference Guidebook, v. 1977, p. 379–405.

Oldaker, P. R., 1991, Hydrogeology of the Fruitland Formation, San Juan Basin, Colorado and New Mexico, *in* Schwochow,

S. D., and others, eds., Coalbed methane of Western America: Denver, Rocky Mountain Association of Geologists, p. 61–66.

Osmond, J. C., 1986, Petroleum geology of the Uinta Mountains–White River uplift, Colorado and Utah, *in* Stone, D. S., ed., New interpretations of northwest Colorado geology: Denver, Rocky Mountain Association of Geologists, p. 213–221.

Palmer, I. D., 1993, Induced stresses due to propped hydraulic fracture in coalbed methane wells: Denver, Society of Petroleum Engineers, Rocky Mountain Region, Low Permeability Reservoirs Symposium, SPE paper no. 25861, p. 221–232.

Petroleum Information, 1993, Rocky Mountain Coalbed Methane Report, v. 4, nos. 2 and 4, variously paginated.

Raymond, A. C., and Murchison, D. G., 1991, Influence of exinitic macerals on the reflectance of vitrinite in Carboniferous sediments of the Midland Valley of Scotland: Fuel, v. 70, p. 155–161.

Robson, S. G., and Stewart, Michael, 1990, Geologic evaluation of the upper part of the Mesaverde Group, northwestern Colorado: U.S. Geological Survey Water-Resources Investigations Report 90-4020, 125 p.

Roehler, H. W., 1965, Summary of pre-Laramide Late Cretaceous sedimentation in the Rock Springs Uplift area, *in* DeVoto, R. H., and Bitter, R. K., eds., Sedimentation of Late Cretaceous and Tertiary outcrops, Rock Springs Uplift: Wyoming Geological Association Guidebook, 19th field conference, p. 11–12.

_________ 1987, Surface-subsurface correlations of the Mesaverde Group and associated Upper Cretaceous Formations, Rock Springs, Wyoming, to Mount Harris, Colorado: U.S. Geological Survey Miscellaneous Field Studies Map MF-1937.

_________ 1988, The Pintail coal bed and Barrier Bar G—a model for coal of barrier bar-lagoon origin, Upper Cretaceous Almond Formation, Rock Springs coal field, Wyoming: U.S. Geological Survey Professional Paper 1398, 60 p.

_________ 1990, Stratigraphy of the Mesaverde Group in the central and eastern Greater Green River Basin, Wyoming, Colorado, and Utah: U.S. Geological Survey Professional Paper 1508, 52 p.

Roehler, H. W., and Hansen, D. E., 1989, Surface–subsurface correlations of the Upper Cretaceous Mesaverde Group and associated formations, Cow Creek in southwest Wyoming to Mount Harris in northwest Colorado: U.S. Geological Survey Miscellaneous Field Studies Map MF-2077.

Ryder, R. T., 1988, Greater Green River Basin, *in* Sloss, L. L., ed., Sedimentary cover—North American craton, U.S.: Geological Society of America, Decade of North American Geology, v. D-2, p. 154–165.

Sales, J. K., 1983, Collapse of Rocky Mountain basement uplifts, *in* Lowell, J. D., ed., Rocky Mountain foreland basins and uplifts: Denver, Rocky Mountain Association of Geologists, p. 79–97.

Schwartzer, R. R., 1983, Variation in the quantity of methane adsorbed by selected coals as a function of coal petrology and coal chemistry: U.S. Department of Energy contract no. DE-AC21-80MC14219, final draft report, variously paginated.

Scott, A. R., 1993, Composition and origin of coalbed gases from selected basins in the United States, *in* Proceedings of the 1993 International Coalbed Methane Symposium, The University of Alabama, School of Mines and Energy Development, Paper 9270, v. 1, p. 174–186.

Scott, A. R., and Ambrose, W. A., 1992, Thermal maturity and coalbed methane potential of the Greater Green River, Piceance, Powder River, and Raton Basins (abs.): American Association of Petroleum Geologists Official Program with Abstracts, Calgary, Alberta, Canada, p. 116.

Scott, A. R., and Kaiser, W. R., 1991, Relation between basin hydrology and Fruitland gas composition, San Juan Basin, Colorado and New Mexico: Quarterly Review of Methane from Coal Seams Technology, v. 9, no. 1, p. 10–18.

Scott, A. R., Kaiser, W. R., and Ayers, W. B., Jr., 1991a, Thermal maturity of Fruitland coal and composition and distribution of Fruitland Formation and Pictured Cliffs gases, *in* Ayers, W. B., Jr., and others, Geologic and hydrologic controls of the occurrence and producibility of coalbed methane, Fruitland Formation, San Juan Basin: Chicago, Gas Research Institute, Topical Report GRI-91/0072, p. 243–270.

_________ 1991b, Composition, distribution, and origin of Fruitland Formation and Pictured Cliffs Sandstone gases, San Juan Basin, Colorado and New Mexico, *in* Schwochow, S. D., Murray, D. K., and Fahy, M. F., eds., Coalbed methane of western North America: Denver, Rocky Mountain Association of Geologists, p. 93–108.

Siepman, B. R., 1986, Facies relationships in Campanian wave-dominated coastal deposits in Sand Wash Basin, *in* Stone, D. S., ed., New interpretations of northwest Colorado geology: Denver, Rocky Mountain Association of Geologists, p. 157–164.

Sklenar, S. E., and Anderson, D. W., 1985, Origin and evolution of an Eocene lake system within the Washakie Basin of southwestern Wyoming, *in* Flores, R. M., and Kaplan, S. S., eds., Cenozoic paleogeography of the west-central United States: Rocky Mountain Section of Economic Paleontologists and Mineralogists, Rocky Mountain Paleogeography Symposium 3, p. 231–245.

Smith, J. B., Ayer, M. F., Knox, C. C., and Pollard, B. C., 1972, Strippable coal reserves of Wyoming: U.S. Bureau of Mines Information Circular IC-8538, 51 p.

Smith, L., Forster, C., and Evans, J., 1990, Interaction of fault zones, fluid flow, and heat transfer at the basin scale, *in* Neuman, S. P., and Neretnieks, I., eds., Hydrology of low-permeability environments: Hanover, Verlag Heinz Heise, International Association of Hydrologists, v. 2, p. 41–67.

Soil Conservation Service, 1983, Wyoming average annual precipitation 1941–1970: U.S. Department of Agriculture, Soil Conservation Service, scale 1:1,000,000.

Stach, E. M., Makowshy, M., Teichmüller, G. H., Taylor, D., Teichmüller, C., and Teichmüller, R., 1982, Stach's textbook of coal petrology (3rd): Berlin, Borntraeger, 535 p.

Stone, D. S., 1975, A dynamic analysis of subsurface structure in northwestern Colorado, *in* Bolyard, D. W., ed., Deep drilling frontiers in the central Rocky Mountains: Denver, Rocky Mountain Association of Geologists, p. 33–40.

Surdam, R. C., and Stanley, K. O., 1980, The stratigraphic and sedimentologic framework of the Green River Formation, Wyoming, *in* Harrison, A., ed., Stratigraphy of Wyoming: Wyoming Geological Association Guidebook, 31st field conference, p. 205–221.

Tang, Y., Jenden, P. D., and Teerman, S. C., 1991,

Thermogenic methane formation in low-rank coals—published models and results from laboratory pyrolysis of lignite, *in* Manning, D. A. C., ed., Organic geochemistry—advances and applications in the natural environment: Manchester, Manchester University Press, p. 329–331.

Teichmüller, M., and Teichmüller, R., 1981, The significance of coalifications studies to geology—a review: Bulletin, Centres Recherche, Exploration-Production, Elf-Aquitaine v. 5, no. 2, p. 491–534.

Thomas, J., Jr., and Damberger, H. H., 1976, Internal surface area, moisture content, and porosity in Illinois coals: variations with coal rank: Illinois State Geological Survey Circular 493, 38 p.

Tóth, J., 1980, Cross-formational gravity-flow of ground-water: a mechanism of the transport and accumulation of petroleum (the generalized hydraulic theory of petroleum migration), *in* Roberts, W. H., III, and Cordell, R. J., eds., Problems of petroleum migration: American Association of Petroleum Geologists, Studies in Geology no. 10, p. 121–167.

Tremain, C. M., and Toomey, J., 1983, Coalbed methane desorption data: Colorado Geological Survey Open-File Report 81-4, 514 p.

Tremain, C. M., Laubach, S. E., and Whitehead, N. H., III, 1991a, Coal fracture (cleat) patterns in Upper Cretaceous Fruitland Formation, San Juan Basin, Colorado and New Mexico—implications for coalbed methane exploration and development, *in* Schwochow, S. D., Murray, D. K., and Fahy, M. F., eds., Coalbed methane of western North America: Denver, Rocky Mountain Association of Geologists, p. 49–59.

__________ 1991b, Coal fracture (cleat) patterns in Upper Cretaceous Fruitland Formation, San Juan Basin, Colorado and New Mexico: implications for coalbed methane exploration and development, *in* Ayers, W. B., Jr., and others, Geologic and hydrologic controls on the occurrence and producibility of coalbed methane, Fruitland Formation, San Juan Basin: The University of Texas at Austin, Bureau of Economic Geology, topical report prepared for the Gas Research Institute under contract no. 5087-214-1544 (GRI-91/0072), p. 97–117.

Tweto, Ogden, 1975, Laramide (Late Cretaceous–early Tertiary) orogeny in the southern Rocky Mountains, *in* Curtis, B. F., ed., Cenozoic history of the southern Rocky Mountains: Geological Society of America Memoir 144, p. 1–44.

__________ 1976, Geologic map of the Craig 1° × 2° quadrangle, northwestern Colorado: U.S. Geological Survey Miscellaneous Investigations Series Map I-972, scale 1:250,000.

__________ 1979, Geologic map of Colorado: U.S. Geological Survey, scale 1:500,000.

__________ 1980, Tectonic history of Colorado, *in* Kent, H. C., and Porter, K. W., eds., Colorado geology: summary of Laramide orogeny in Colorado: Denver, Rocky Mountain Association of Geologists Symposium, p. 129–134.

Tyler, Roger, Ambrose, W. A., Scott, A. R., and Kaiser, W. R., 1991, Coalbed methane potential of the Greater Green River, Piceance, Powder River, and Raton Basins: The University of Texas at Austin, Bureau of Economic Geology, topical report prepared for the Gas Research Institute under contract no. 5087-214-1544 (GRI-91/0315), 248 p.

__________ 1992a, Evaluation of the coalbed methane potential in the Greater Green River, Piceance, Power River, and Raton Basins, *in* Mullens, C. E., ed., Rediscover the Rockies: Wyoming Geological Association Guidebook, 45th field conference, p. 269–302.

Tyler, Roger, Kaiser, W. R., Ambrose, W. A., Scott, A. R., Laubach, S. E., and Ayers, W. B., Jr., 1992b, Coalbed methane characteristics in the foreland of the Cordilleran thrust belt, western United States, *in* Beamish, B. B., and Gamson, P. D., eds., Symposium on coalbed methane research and development in Australia: James Cook University of North Queensland (Coalseam Gas Research Institute), v. 2, p. 11–32.

Tyler, Roger, Laubach, S. E., Ambrose, W. A., Grout, M. A., and Tremain, C. M., 1992c, Face-cleat patterns in Rocky Mountain foreland basins, western United States: permeability indicators for coalbed methane (abs.): American Association of Petroleum Geologists Bulletin, v. 76, no. 8, p. 1269–1270.

Tyler, Roger, Laubach, S. E., and Ambrose, W. A., 1991, Effects of compaction on cleat characteristics: preliminary observations, *in* Ayers, W. B., Jr., and others, Geologic and hydrologic controls on the occurrence and producibility of coalbed methane, Fruitland Formation, San Juan Basin: Chicago, Gas Research Instutute Topical Report GRI-91/0072, p. 141–150.

Verbeek, E. R., and Grout, M. A., 1986, Cenozoic stress rotation, northeastern Colorado Plateau, *in* Stone, D. S., and Johnson, K. S., eds., New interpretations of northwest Colorado geology: Denver, Rocky Mountain Association of Geologists, p. 97.

Weimer, R. J., 1965, Stratigraphy and petroleum occurrences, Almond and Lewis Formations (Upper Cretaceous), Wamsutter Arch, Wyoming, *in* DeVoto, R. H., and Bitter, R. K., eds., Sedimentation of Late Cretaceous and Tertiary outcrops, Rock Springs Uplift, Wyoming: Wyoming Geological Association Guidebook, 19th field conference, p. 65–80.

Zoback, M. L., and Zoback, M. D., 1989, Tectonic stress field of the continental United States, *in* Pakiser, L. C., and Mooney, W. D., eds., Geophysical framework of the continental United States: Geological Society of America Memoir 172, p. 523–539.